T0320955

Insect Phenotypic Plasticity

Diversity of Responses

Insect Phenotypic Plasticity

Diversity of Responses

Editors

T.N. Ananthakrishnan
Formerly Director of
Entomology Research Institute
Chennai, India

Douglas Whitman
Department of Biological Sciences
Illinois State University
Normal, IL, USA

Science Publishers, Inc.

Enfield (NH), USA   Plymouth, UK

SCIENCE PUBLISHERS, INC.
Post Office Box 699
Enfield, New Hampshire 03748
United States of America

Internet site: *http://www.scipub.net*

sales@scipub.net (marketing department)
editor@scipub.net (editorial department)
info@scipub.net (for all other enquiries)

Library of Congress Cataloging-in-Publication Data

Insect phenotypic plasticity editors, T.N. Ananthakrishnan, Douglas Whitman.
   p. cm.
  Includes bibliographical references and index.
  ISBN 1-57808-322-2
   1. Insects--Ecology. 2. Insects--Genetics. 3. Phenotype.   I. Ananthakrishnan, T.N.,
1925-II. Whitman, Douglas (Douglas W.)

  QL496.4.I565 2005
  595.713'5--dc22

2004065384

ISBN 1-57808-322-2

Published by Science Publishers, Inc., Enfield, NH, USA
Printed in India.

Preface

Evolution has long been thought to occur primarily as a result of natural selection acting on existing genetic variation. However, this model is misleading, because natural selection acts not on genes, but on phenotypes. This is a critical concept, because phenotypes are determined, in part, by the environment. Indeed, all organisms exhibit phenotypic plasticity—the expression of different phenotypes in a single genotype when subjected to different environments. Environments influence organismal development inducing different, and often permanent, developmental outcomes. Organisms can also respond immediately to environmental factors, with rapid and sometimes reversible changes in behavior, physiology, morphology, and life history. These short- or long-term environmentally induced changes can have profound consequences for fitness. Hence, the environment serves a dual role in evolution: it may both *generate* phenotypic variation, and *select* among that variation. Thus, phenotypic plasticity plays an important role in evolution.

In *Insect Phenotypic Plasticity*, we document the plasticity inherent in insects. Phenotypically plastic traits, include morphological, behavioral, and physiological, characteristics. These environmentally induced differences can serve as the raw products upon which natural selection acts.

Phenotypic plasticity, in short, deserves increased attention by those involved in studies on biological diversity and is of practical concern for agricultural and medical entomology.

T.N. Ananthakrishnan
Chennai, India
ananthrips@yahoo.com

Douglas W. Whitman
Normal, IL USA
dwwhitm@ilstu.edu

Contents

1

Perspectives and Dimensions of Phenotypic Plasticity in Insects

T.N. Ananthakrishnan
Flat No. 6, Dwarka, 42 (Old No. 22), Kamdar Nagar, Nungambakkam, Chennai 600034, India; e-mail: ananththrips@yahoo.com

Introduction

All species exhibit phenotypic variation. In some cases, this variation is directly controlled by underlying genetic variation, whereas in other cases, variation in phenotype is induced by the environment. The latter case is termed phenotypic plasticity, and is defined here as the expression of different phenotypes in a single genotype when subjected to different environments. Phenotypic plasticity is a complicated subject that has recently received a tremendous amount of attention, largely because of the realization that phenotypic plasticity is ubiquitous, and is important for understanding the ecology, taxonomy, and evolution of populations and species (Scheiner 1993, Nylin and Gotthard 1998, Schlichting and Pigliucci 1998, Pigliucci 1996, 2001, Agrawal 2001, West-Eberhard 2003). A knowledge of the patterns of such plasticities and an assessment of their implication are useful in the proper assessment of species.

Understanding phenotypic plasticity requires comprehension of how genes and the environment interact to produce the myriad of traits that comprise the whole organism. Genes code for gene products and the regulation of their expression. Starting with incipient development, translation of genetic information creates structural proteins and enzymes, which produce enzymatic products. However, these products immediately begin to interact with one another in epigenetic developmental cascades that in time move further away from direct genetic control. Gene products create a cellular environment that can be influenced by external factors, including heat, light, nutrition, pH, toxins, pathogens, etc. In some cases, the environment can directly influence gene expression. But more often externally induced changes in the internal environments of developing organisms alter their development, physiology, or behaviour, producing different phenotypes. Hence, the genotype does not give rise to the phenotype, but to a *range of phenotypes*. In other words, "a phenotype is a biological system constructed by successive interactions of the individual genotype with the environment, in which development takes place. The norm of the reaction (Schlichting & Pigliucci, 1998) is the entire range, the whole repertoire of the variant pathways in development, that may occur in the carrier of a given genotype in all environments, favourable or unfavourable, natural or artificial" (Dobzhansky, 1971).

Thus, phenotypes are always the product of genes x the environment. Indeed, genes can never be separated from the environment, because even the chromosomes of an unfertilized egg reside in a cellular environment that was presumably influenced by the mother's physiological state, and continues to be influenced by prevailing environmental conditions.

This leads to the realization that most traits or characteristics of organisms are plastic, i.e., they can be influenced by the environment. It also suggests that phenotypic plasticity functions in evolution, by initiating phenotypic variation, the raw product upon which natural selection acts. Natural selection selects among different phenotypes, not genes, and variation is an indispensable prerequisite of evolution. When the environment induces a novel phenotype, then natural selection can begin to act on that novelty. If that specific phenotype never occurs, then it cannot be selected for. Phenotypic plasticity then comes to assume an important place in evolution. Thus, understanding the cause and consequence of phenotypic variation is important for understanding the mechanisms of evolution.

Phenotypic plasticity is also important in ecology and ecological success. Environments are variable. When faced with changing conditions, individuals have greater success when they can produce adaptive plastic responses. Phenotypic plasticity therefore can serve as a buffering mechanism against environmental variation (Schlichling and Lewis, 1986). Indeed, phenotypic plasticity may allow a species to greatly enlarge its ecological niche, which again, feeds back to greater opportunities for selection, diversification and radiation. Phenotypic plasticity may be important in species-species interactions such as symbiosis, mutualism, parasitization, predation, etc., as each partner reciprocates to the other with higher levels of plastic response. Agrawal (2001) uses the term 'reciprocal phenotypic change' for the "escalation of phenotypes between *individuals of two species* with a developmental change in the phenotype". This concept is similar to that of coevolution, except that coevolution acts in populations over evolutionary time, whereas reciprocal phenotypic change acts in real time in individuals. In the words of Thompson (1994) "evolution of adaptation within species becomes intertwined with the evolution of interactions." Hence, phenotypic plasticity can enable maximum fitness in variable environments - an aspect Agrawal (2000) refers to as the 'adaptive plasticity hypothesis'.

Phenotypic plasticity also impacts our understanding of taxonomy, because it suggests that species characteristics are not immutable, but are influenced by the environment and can be highly variable. Indeed the entomological literature is replete with cases of mistaken identity of species due to non-recognition of variation within populations of a species. One of the most spectacular of these relate to the thrips, *Ecacanthothrip tibialis* Ashmead, wherein populations studies revealed the presence of more than twenty synonyms, especially due to considerable disparity between different forms evident in the males (Ananthakrishnan, 1969). Phenotypic plasticity also impacts basic research and medical and agricultural entomology, not only when it inhibits our ability to distinguish among species, races, or populations of insects, but when plastic responses of pests thwart our attempts to control them. In contrast, the adaptive, plastic responses of natural enemies to their prey can aid biological control. The consequences of phenotypic plasticity therefore are diverse and an understanding of the functional impact of plasticity plays an important role leading to ecological and evolutionary success of organisms, and our ability to understand and manipulate them.

In terms of traits that can be influenced by the environment, phenotypic plasticity acts on a wide range of biological characteristics that sorts broadly into biochemistry, morphology, physiology, and behaviour, but includes such diverse phenomena as colour polymorphisms, castes in social insects, migration, phase change in locusts, allometry, immune response, enzyme induction, induced defenses, diapause, quiescence, aestivation, freeze tolerance, heat shock, imprinting, learning, life history tradeoffs, environmental constraints, canalization, physiological tradeoffs and homeostasis, and maternal effects (Mousseau and Fox 1998, Nylin and Gotthard 1998, Emlen and Nijhout 2000, West Eberhardt, 2003). As such, phenotypic plasticity invades virtually all disciplines of entomology.

Note, that phenotypic plasticity ascribes no fitness value to the altered phenotype: an environmentally induced change in phenotype may or may not be beneficial or adaptive. The resulting plastic response may simply be a non-adaptive constraint imposed by a specific environmental factor, such as when poor nutrition, high temperature, toxins, or disease during larval development results in a smaller adult insect. In such cases, the organism is said to be "susceptible" to the environmental factor. In other cases of phenotypic plasticity, the plastic response may be beneficial. For example, some caterpillars and grasshoppers develop larger heads and mandibular muscles when fed tough or fibrous food (Thompson, 1992). Presumably the environmentally induced plastic response is advantageous because it allows the insect to better feed on a tough diet. Other cases of phenotypic plasticity are clearly adaptive, highly evolved, and under genetic control for the plastic response. Included are some of the most intriguing phenomena known to entomologists, such as polymorphism in aphids, locusts, butterflies, and social insects. Some plastic responses, such as diapause induction in insects are "anticipatory", and rely on highly evolved sensory and physiological mechanism that monitor predictive environmental cues and translate them into appropriate, adaptive physiological responses, prior to the occurrence of the harmful factor (Tauber *et al.*, 1986).

Hence, phenotypic plasticity has to be regarded not merely as intrinsic variation, but as a fundamental character of insects, subject to evolutionary pressures, and of practical concern for taxonomy and agricultural and medical entomology. Thus, phenotypic plasticity deserves increased attention by those involved in studies on biological diversity wherein phenotypic plasticity is an integral component. In the following are highlighted specific examples of phenotypic plasticity in certain groups of insects. Needless to emphasize that this includes some striking instances of polymorphism in mycophagous and gall inhabiting thrips, which show an unbelievably large range of structural, behavioural, and reproductive polymorphism, an aspect dealt with more exhaustively by Mound in chapter 4 of this volume. The multifold dimensions of phenotypic plasticity discussed here serve as a brief introduction to the subject.

Grasshoppers and Locusts

Grasshoppers exhibit phenotypic plasticity in a wide range of traits. Among the most striking and best studied is phase change in certain species of locust, wherein high population densities cause solitary morphs to change into gregarious morphs (Uvarov 1966, 1977). Although the mechanism is still not completely understood, it is believed that high levels of a "gregarization pheromone' build up in dense populations and trigger hormonal changes that stimulate the transformation from solitary to gregarious forms (Loher 1990, Dale and Tobe 1990). Gregarious locusts aggregate and move in swarms of hundreds of thousands of individuals that fly great distances and eat ravenously. Gregarious phase *Locusta migratoria* (Linn.) and *Schistocerca gregaria* Forskal tend to be smaller, with fewer ovarioles, with small egg clutches and with larger eggs than solitary forms (Uvarov 1966). In *Locustana pardalina* (Walker), gregarious individuals are larger and lay larger clutches than solitary individuals (Uvarov 1966). Gregarious locusts also differ in colour from solitary locusts. Solitary *L. migratoria* hoppers are green, while the gregarious ones show various combinations of black, yellow, red, etc. The phenomenon of locust phases therefore encompasses intraspecific diversity in morphology, physiology, and behaviour, all of which affects the locust's life history and ecology.

Grasshoppers exhibit plasticity in other traits as well, including diapause (Uvarov 1966; Lee and Whitman 2005) and sexual strategies (Greenfield and Shelly 1985, 1990). Many species are highly variable in clutch size or pod interval, laying smaller clutches when stressed (Stauffer and Whitman, 1997), and in a few species, females show plasticity in egg size, older females laying larger eggs (Stauffer and Whitman 1997). The number of moults is variable in some taxa, with young larvae exposed to cold, skipping one or two moults, reaching adulthood earlier (Dean 1982, Dingle *et al.*, 1990). Larvae from small eggs undergo additional moults, resulting in larger adults (Grant *et al.*, 1993). Brachypterous females mate and oviposit earlier and faster, and are more fecund than the long winged form (Ritchie *et al.*, 1987).

Lepidoptera

Several Lepidoptera species exhibit phases, including *Laphygma, Plusia,* and many sphingids and noctuids. In *Spodoptera exempta,* which shows density-dependent polyphenism, solitary forms are green and highly cryptic in their behaviour, whereas the gregarious larvae are conspicuously black and feed voraciously. Gregarious caterpillars grow faster, perhaps due to greater ability to thermoregulate, as a result of their black color (Rhoades, 1985). Crowded, gregarious forms have a larval duration of 16 days as compared to 24 for the solitary forms. The gypsy moth, *Lymantria dispar* shows large changes in activity patterns and behaviour in response to temperature, along a broad continum, from strict 'thermoconformers' to more or less regular "thermoregulators" (Casey, 1993). Caterpillars differ in thermal sensitivity and thermal optima of important processes such as growth, metabolism and feeding rate. Thermoconformers are sluggish. One of the best examples of phenotypic plasticity, related to thermal responses is shown by *Spodoptera exempta,* which exhibits density-dependent polyphenism, and different growth rates by solitary and gregarious phases at constant temperatures. Crowded caterpillars have a larval duration of 16 days, as against 24 for solitary forms under the same temperature.

Photoperiod often serves as an important cue for seasonally dimorphic Lepidoptera; a difference of a few hours affects the pattern of development. One of the oldest examples is the butterfly *Araschnia levana,* whose summer form was named *Araschnia prorsa.* It was later discovered that when caterpillars are exposed to long day length of over 16 hours per day, the darker form, *A. prorsa,* resulted, and with the exposure for only 12 hours, light-coloured forms were evident (Beck, 1963).

Webworm moths such as *Hyphantria cunea* have three generations with the third generation having the largest body size, and the second generation, the smallest. The differences in body sizes of these three generations influence their foraging behaviour and susceptibility to natural enemies. Such differing adult sizes have been reported also for some geometrid species, the overwintering larvae giving rise to larger adults than those of the summer generation (Schweitzer, 1977).

Membracids

Many membracids show high degrees of morphological plasticity, especially in the size of their suprahumeral horns. Indeed, there are so many overlapping character states in adult membracids of different species that it causes taxonomic problems, requiring the study of nymphs for species identification. This is seen in species of *Tricentrus* Stal, *Leptocentrus* Distant, *Oxyrhachis* Germar, *Enchinopa,* and *Otinotus* Buckton. A species of *Tricentrus* lacking suprahumeral horns is liable to be mistaken for a species of *Gargara*. That the suprahumeral horns are likely to vary considerably within species is clearly exemplified by *Oxyrachis tarandus* (Fabricius) and *Leptocentrus variicornis* Ananthasubramanian and Ananthakrishnan. In *Tricentrus pilosus* Ananthasubramanian and Ananthakrishnan for instance, the supra-humeral horns in females show four different types - with normal horns, short horns, aborted horns and those without horns, while the males show only three types (Ananthasubramanian and Ananthakrishnan, 1975).

Whiteflies

Aleyrodids also show considerable intraspecific diversity in pupal morphology, often induced by the physical nature of the host leaf, as seen in

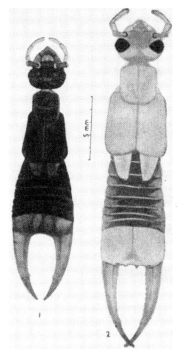

Fig. 1   *Labidura riparia* showing colour r size variants

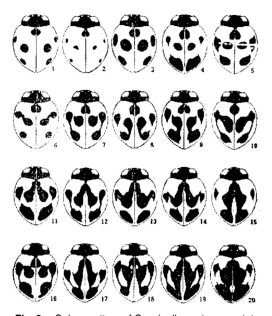

Fig. 2   Colour pattern of *Coccinella septempunctata*

Trialeurodes rara Singh and *Bemisia tabaci* (Gennadius). *T. rara* shows this tendency to a remarkable degree, in that in those forms infesting pubescent leaves, setae are well developed, as against the absence of setae when infesting glabrous hosts (David and Ananthakrishnan, 1984).

Dermaptera

Among the earwigs, macrolabic, cyclolabic and microlabic forms exist, showing large-scale variations in their forceps, besides diversities relating to apterous, brachyterous and macropterous forms. These different phenotypes are thought to result from nutritional and temperature differences during development. In any given species, the forceps may be strongly incurved with smooth internal margins with a basal verticular triangular tooth, the apterous form with less strongly curved forceps tapering apically with the tip pointed and gently hooked.

Coleoptera

Many stag beetles, scarabaeids, curculionids and other Coleoptera show size plasticity and allometry, often related to nutrition. Work on the nature of sexual selection in the brentid weevil (*Brentus anchorago*), has shown impressive phenotypic variation, the more elongate males in particular showing considerable diversity in the long, stronger rostrum and the powerful grossly enlarged mandibles playing a significant role in mating success (Johnson, 1982).

Aphids

Many aphids produce parthenogenetic young under the influence of long day length, and oviparous forms when days are short. For example, in *Megoura viciae* sexually reproducing forms are produced when grown in photoperiod of less than 14.5 hours of light per day. When it is more than 15 hours of light, non-sexual forms are produced. Photoperiod exerts its effects by influencing hormone production in the developing aphid nymph (Beck, 1963).

Thysanoptera

Sizeable populations of mycophagous and gall thrips exhibit considerable diversity, wherein the extreme forms of individuals are almost unrecognizable. The related variations include alary polymorphism and polychromatism, the winged and apterous individuals showing considerable variation; besides variations of a profound nature mostly confined to the males, sometimes in females, wherein bizarre, grotesque males have certain parts strikingly enlarged *(oedymerous)*, while at the other end, the individuals are bereft of such anomalies *(gynaecoids)*. To distinguish from the males, the females, which show such extreme diversities, are termed major and minor females. Interestingly enough, new characters are expressed in the oedymerous males and major females involving cephalic, thoracic, abdominal horns, with excessively enlarged, armed prothorax and fore femora, with a wholly new array of setae, which are more prominent in the apterous forms. While Ananthakrishnan (1961, 1967, 1969, 1970, 1984, 1990) has provided several instances of such diversities, Mound (2003) in this volume provides an analytical and meaningful review of all aspects of the phenotypic plasticity of thrips. Instances provided here further supplement the structural, behavioural and reproductive aspects of thrips polymorphism.

The gynaecoid males by virtue of their minimal organ development resemble the females in general make-up. Studies by Ananthakrishnan on species of *Ecacanthothrips* Bagnall, *Hoplandrothrips* Hood, *Kleothrips* Schmutz, *Elaphrothrips* Buffa, *Nesothrips* Kirkaldy and *Tiarothrips* Priesner, to mention a few, indicate the trends involved in the transition from the gynaecoid through the normal male to the oedymerous male. The gynaecoid and the maximum oedymerous forms being at the two ends of a series are totally different in general make-up. Patterns of intraspecific diversity exist, those with minimal effects on the morphs being termed as *monophasic* or *simple*, while those which show excessive development of traits are referred to as *multiple* or *polyphasic*. Multiple or polyphasic patterns therefore involve not only pronounced development of several parts and varying with species or species groups, but also result in the development of certain additional structures only in the extreme oedymerous individuals and not known in the normal males. In species of *Ecacanthothrips* the oedymerous males develop very strong forefemora with 2 or 3 subapical teeth, a basal or apical foretibial teeth, longer prothoracic setae and a strong pronotum. The oedymerous males have well developed coxal prolongation and the outer margin of the fore femora at base tends to be clearly concave, fringed with a cluster or long hairs. This concavity becomes progressively reduced, along with the size and number of the fringing hairs as we proceed down the series to the gynaecoid (Ananthkrishnan, 1961, 1970) (Figs. 3 and 4).

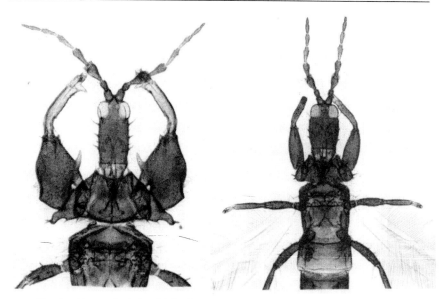

Fig. 3   *Ecacanthothrips tibialis*. Oedymerous male (left) gynaecoid male (right)

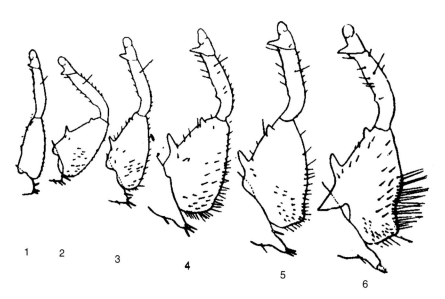

Fig. 4   The series from gynaecoid (1) to oedymerons (4-6) male foreleg

Patterns recognizable in *Tiarothrips subramanii* (Ramakrishna) vary from individuals with short head production, simple, non-corrugate, weakly setose third antennal segment, to individuals with excessively elongate head production and very long third antennal segment, strongly and asymmetrically sinuate, with deep concavities and strong apical armature. The individuals from the two extremes are so strikingly different that they could well represent two different species taking the traditional standards of morphometry (Ananthakrishnan, 1973) (Figs. 5 and 6). Equally spectacular is *Nesothrips falcatus* Ananthakrishnan, where the oedymerous males develop highly elongate and broadened forefemora, forecoxa and forefemora with very strong, curved chitinous hook-like structures, mid-coxae also with a chitinous tooth excessively long as compared with the gynaecoid (Fig.7).

Fig. 5   Aggregation of *Tiarothrips*

Gall inhabiting Tubulifera also exhibit oedymery and gynaecoidy to a remarkable degree. It is mostly the females, which show enlarged forefemora and foretarsal armature as in species of *Arrhenothrips* Hood, *Mallothrips* Ramakrishna (Fig.6). Some of these robust built individuals function as soldiers in the defense of a gall. Intragall dimorphism of thrips together with intergall variation might be related to different degrees of wing polymorphism as well as to natural interpopulation variation (Ananthakrishnan and Raman, 1989; Mound, 1994). Mound (1995) and Crespi (1992) also indicate that inter-gall variation together with intra-gall

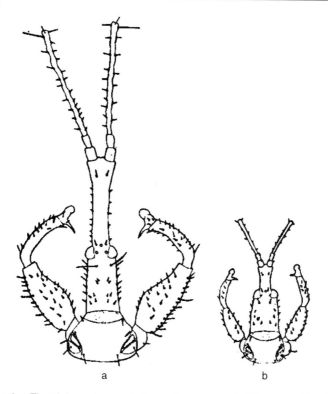

Fig. 6   *Tiarothrips subramanii*—(a) oedymerous male (b) gynaecoid male

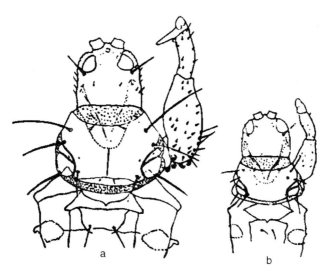

Fig. 7   *Nesothrips falcatus*—(a) oedymerous male (b) gynaecoid male

dimorphism might be related to different degrees of microptery. The extraordinary diversity shown by thrips species inhabiting the woody galls on casuarina trees in Australia is something very unusual, with species of *Iotatubothrips* Mound, *Phallothrips* Mound and *Thaumatothrips* Karny exhibiting unusually high intrapopulation variation within all the morphs.

An interesting aspect of alary polymorphism is seen in the Acacia gall thrips, *Thilakothrips babuli* Ramakrishna, wherein with leaf fall, apterous adults undergo diapause and with acacia sprouting flower buds, they leave the diapause site, migrating to the inflorescence to form floral galls within which both macropterous and apterous forms develop. Gall dimorphism is equally evident in the aphid gall on *Populus*, made by *Pemphigus populitransversus* where early galls are elongate and later galls are globular (Wool, 1982).

Functional convergence in spatially limited bark-infesting thrips species leads to gregariousness, inbreeding, and overlapping of generations tending towards prosociality (Crespi, 1986, 1990). A higher level of organization bordering on parasocial behaviour is seen in *Anactinothrips gustaviae* Hood. Resource partitioning coupled with intra- and interspecific competition for mating and egg laying is viewed as an important strategy in which intraspecific variation results from polymorphism. The diversity is extremely vast in the case of mycophagous forms. Exaggerated size attributable to successful mate selection as a sex-limited structural and behavioural polymorphic phenomenon, contributes to the reproductive success of especially the oedymerous tubuliferan males. The behavioural strategy of defense polygamy, wherein dominant males mate with many females, is typical of the mycophagous species, *Hoplothrips pedicularius*. While dominant males show stabbing and wagging behaviour during fighting, sneaking behaviour is typical of subordinate males (Crespi, 1986). For example, the males of *Hoplothrips pedicularius* and *Tiarothrips subramanii* that guard the oviposition sites are known to pursue a "hawk" strategy. Dominance hierarchy is seen in several mycophagous species, wherein the dominant males disable the rivals and keep them from mating, thus defending their females. Low relatedness of males and defendable resources in the absence of alternate strategies leads to fighting between males in female aggregation. The development of varied patterns of sex-limited polymorphism introduces complications in mating behaviour and parental care that tend to be different in the extreme individuals (Ananthakrishnan, 1987; Crespi, 1992).

Both sexual and parthenogenetic means of reproduction are known among thrips. Oviparity, ovoviviparity, and viviparity are especially

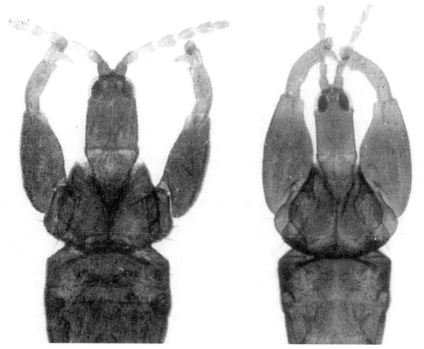

Fig. 8 *Arrhenothrips ramakrishnae*; (left) minor female; (right) major female

common among the mycophagous forms, while arrhenotoky and thelytoky are restricted particularly to phytophagous species. Phenotypic changes alter the fitness of the genotype in terms of successful sperm production and oviposition mediated through mating patterns. The major females are more fecund and mate more frequently, but selectively, with the oedymeres. Large egg size demands proportionately prolonged developmental durations and consequently delayed reproduction (Ananthakrishnan, 1990). Fecundity, oviposition rate and egg laying sequence in mating types of the mimusops gall thrips, *Arrhenothrips ramakrishnae* Hood offer good examples of plasticity (Varadarasan and Ananthakrishnan, 1981).

Winglessness, which is more often correlated with oedymery as in *Brithothrips fuscus* Mound and *Hoplothrips* species, is frequently associated with the males derived through regular inbreeding that results from gregariousness and haplodiploidy. Gyandromorphs and intersexes are on record in *Oxythrips flavus* and *Thrips fuscipennis* (Mound, 1970).

Ovarian polymorphism is well evident in species of *Bactrothrips, Tiarothrips, Meiothrips* and *Elaphrothrips*, to mention a few, associated with ovipary, ovovivipary and vivipary. Both the ovoviviparous and viviparous ovaries differ from each other, as well as from the oviparous ovaries in structure and function. The oviparous ovarioles are longer due to the long vitellogenic zone with the vitellogenic zone containing larger basal oocytes. The viviparous ovarioles are shorter due to the absence of vitellarium. In the ovoviviparous ovarioles the vitellarium is short, but with larger basal oocytes. As a result of structural variation among polymorphic ovaries, their relative positions are also considerably altered (Dhileepan and Ananthakrishnan, 1987) (Fig.9).

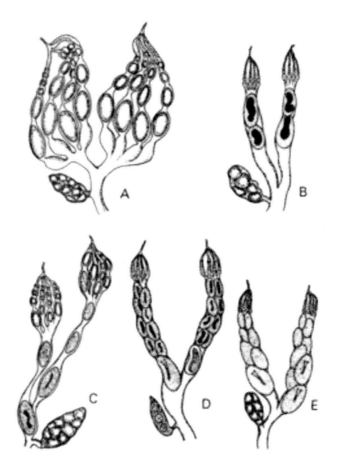

Fig. 9   Ovarian types in *Tiarothrips subramanii* (A) Oviparous; (B, C) Partially oviparous; (D) Ovoviviparous; (E) Viviparous

Herbivorous Insects and Plants

Chemical ecology has contributed substantially to our understanding of the evolution of phenotypic plasticity. Herbivores have altered the "expression of many traits over evolutionary time, but can also alter the expression of those traits in ecological time" (Karban and Myers, 1989; Tallamy and Raupp, 1991; Karban and Baldwin, 1997). Baldwin (1999) has indicated that, "the environment plays two roles in the life of every organism: (a) as a provider of signals that organisms use to modify their phenotypes (b) as a provider of fitness function that selects those phenotypes that are most fit for a particular environment, so that phenotypic plasticity is a major evolutionary response to environmental variation". For instance for caterpillars, large size could buffer against physiological stress, perhaps a reason as to why generalists tend to be larger than specialists. Further, gregarious caterpillars through their mass attack can overwhelm host plant defenses unlike the solitary ones. As such, functional, physiological or behavioural phenotypes become important and instances where even within a population both generalist and specialist phenotypes are maintained are exemplified by the butterfly *Euphrydas editha* (Thomson, 1994). Evidence for preference-performance-correlation has been shown by offspring of specialist females surviving better on plants chosen by their mother than, on rejected plants. On the other hand, offspring from generalist females performed equally well on both categories of plants (Simms, 1998). Resource texture therefore is an aspect to which insects respond in diverse ways (Bernays and Chapman, 1987). In a mixed crop situation, a specialist used to host-specific cues, might be confused or repelled by a nearby non-host species, while a polyphagous generalist species may perceive diverse plant mixtures without adverse effects. So patterns of variation in host chemistry and insect responses to those baits are critically important.

For instance, herbivores may exhibit considerable diversity in response to iridoid; glycosides, the generalist surviving less well on diets with iridoids, and the specialist performing best on diets with two iridoid glycosides in *Plantago lanceolata* (Bower and Puttick, 1988). Secondary compounds also influence the development of chemoreceptors on antennae in acridids. In an odour-free environment, acridids such as *Schistocerca* have fewer basiconic and coeloconic sensilla, but more when exposed to secondary compounds (Chapman and Lee, 1991). Failure to recognize the existence of specialist-generalist variation within a population can result in inaccurate estimates of selection in insect-plant systems.

Parasitic Wasps

Natural enemies are plastic in their behaviour, and foraging behaviour can be continuously modified according to environmental circumstances (Vet and Dicke, 1992). "The eventual foraging effectiveness of a natural enemy's net intrinsic condition is matched with the foraging environment in which it operates" (Lewis *et al.*, 1990). Often naive, laboratory-reared parasites are less effective in locating and attacking hosts than are experienced, field-collected wasps (Whitman 1988). This is seen in *Trichogramma chilonis* emerging from parasitized *Corycra cephalonica* eggs. The need for a better appreciation of behavioural dynamics of parasites cannot be overemphasized, since natural enemies face changing situations in their attempts to find a host, in particular the presence of strong, chemical, visual, auditory cues in relation to the presence of hosts. Likewise, increased hunger may alter search behaviour and host acceptance. As such a pre-release exposure to important stimuli tends to improve the responses of natural enemies through associative learning (Whitman 1988).

Further, parasites are known to adjust their behaviour to changing availability and quality of hosts with different responses evident to odours from plants with different host densities. Experience assists in estimating the informational value of differences in volatile cues between populations. Generally, behaviours of generalists tend to be more flexible than those of specialists (Vet & Dicke, 1992). This is exemplified by the larval endoparasitoid, *Cotesia glomerata*, a generalist, which attacks several pierid hosts on crucifers. This wasp assesses sites based on infochemical qualities, and adjusts its behaviour to spend more time at attractive sites (Geervilet, *et al.*, 1998).

Lewis and Sheehan (1997) have developed a general conceptual model to collectively assess three sources of intrinsic, intraspecific variation in parasitoid foraging, taking into consideration genotypic diversity, influence of different physiological status on responses by individuals and plasticity of individual parasites caused by pre-adult and adult experiences. Emphasis is made of genotypic and phenotypic behavioural traits with the type of environment in which they forage.

Discussion

Mayr (1988) in his book *Toward the New philosophy of Biology* has expressed the view that, "the species plays such different roles in the thinking of various kinds of biologists we must be aware of the differences in the interests and approaches of systematics, evolutionary biologists, ecologists, behavioural biologists, biogeographers and many other kinds of biological scientists". Though it is usually believed that phenotypic plasticity is of great interest to the ecologist and evolutionary biologist, the basic importance to the taxonomists cannot be overlooked, as it is from here that ecology and evolutionary biology arises. While literature abounds in theoretical discussion on phenotypic plasticity, some especially in relation to plant biology, an understanding of the distinctive ways in which new phenotypes are formed in insects, in relation to environment cues as altitudes, temperature and humidity, crowding and the like cannot be ignored. The capacity to exhibit phenotypic variation may be an inescapable consequence of a complex genome and complex physiological pathways, and environment experienced by a population, which determine the phenotypes and establish their fitness for a particular trait (Nanjundiah, 2003).

Maintaining the sensory and regulatory mechanisms needed for plasticity is an important aspect calling for more inputs in relation to the costs and limits of phenotypic plasticity (Dewitt, Sih and Wilson, 1998).

Aspects related to the diversity of approaches to phenotypic plasticity have been briefly indicated, keeping in mind the fact that adaptive plasticity evolved upon the interaction of populations to varied environments. Fluctuations are confronted by populations through phenotypic variations affecting single individuals, within population or in future generations so as to result in different levels of adaptive variation (Meyers and Bull, 2002). The genetic basis of plasticity is an equally important aspect in that some genes may be turned on and of in particular environments and some alleles may be expressed in several different environments. Developmental switches and reaction norms play an important role in phenotypic plasticity. Sometimes an organism produces a phenotype that varies as a continuous function of the environmental signal, so that within each phenotype, the morphology, life history and physiology are integrated to function in a specific ecological role (Stearns, 1989). For instance, the impact of individual variations on the reproductive capacity of an insect or the genetic release mechanisms involved in the production of diverse phenotypes appear important considerations in the reproductive biology of a species. Instances are on record where fitness is increased when reproduction is curtailed and senescence rate is slowed - an aspect called phenotypic plasticity of senescence (Tatar and Yin, 2001). Whenever the structural, physiological and developmental strategies enable prolonged survival of individuals, superior mating preferences are promoted. Mycophagous Tubuliferan thrips have adequately demonstrated the superior forms of the oedymerous males with respect to reproductive efficiency, the increased number of morphs produced becoming better adapted to exploit the environment.

As to the quantum of adaptability or otherwise of such phenotypic responses to environmental variations, Baldwin's (1999) view that "organisms are not infinitely plastic and that phenotypic responses to environmental variations are not frequently adaptive"; a phenotype induced by a particular environment may have higher fitness in that environment than alternative phenotypes. This is a fitness benefit, as against lower fitness in other environments, referred to as benefits of plasticity (Relyea, 2002). The ability of different groups of insects to exhibit structural, functional, reproductive and developmental plasticities is considerable, adaptive in many cases and hence concentrated efforts are needed to identify and coordinate the large-scale occurrence of phenotypic plasticity in several groups of insects which may undoubtedly offer a greater understanding of evolutionary biology.

Acknowledgements

I wish to thank Prof.Vidyanand Nanjundiah of the Indian Institute of Science, Bangalore and Douglas Whitman for their critical observations and suggestions.

References

Agrawal, A.E. 2001 Phenotypic plasticity in the interaction and evolution of species. *Science*, **294**: 321-326.

Agarwala, B. K., Majumdar, K. and Sinha, S. 2002 Electrophoretic variation in the Isoenzymatic pattern of *Lipaphis erysimi* (Kaltenbach) (Homoptera: Aphididae) in relation to host plants and morphs. *Entomon*, **27**: 1-10.

Ananthakrishnan, T.N 1961 Allometry and speciation in *Ecacanthothrips* Bagnall. *Proc. Biol. Soc.Wash.*, **74**: 275-280.

Ananthakrishnan, T.N 1964 A contribution to our knowledge of Tubulifera (Thysanoptera) from India. *Opscula Ent.Suppl.*, **25**: 1-120.

Ananthakrishnan, T.N. 1967 Patterns of structural diversity in the males of some Phlaeophilous Tubulifera (Thysanoptera). *Ann. Soc. Ent. Fr. (N.S)* **4**: 413-418.

Ananthakrishnan, T.N. 1969 Indian Thysanoptera, CSIR Zoological Monograph, **1**: 102-117.

Ananthakrishnan, T.N. 1970 Biosystematics of Thysanoptera. *Ann. Rev. Entomol.* **24**: 159-183.

Ananthakrishnan, T.N. 1970 Trends in intraspecific sex-limited variations in some mycophagous Tubulifera. *J. Bombay Nat. Hist. Soc.* **67**: 481-501.

Ananthakrishnan, T.N. 1973 Mycophagous Tubulifera of India. Occ. Publ. 2, Ent. Res. Unit, Loyola College, 144pp.

Ananthakrishnan, T.N. 1978 Thrips gall and gall thrips. *Zool. Surv. India, Tech. Mon*; 1, 69pp.

Ananthakrishnan, T.N. 1984 *Bioecology of Thrips.* Indira Publishing House, Michigan, 283 pp.

Ananthakrishnan, T.N. 1987 Implications of polymorphism in Thysanoptera. *Insect Sci. appl.* **8**: 435-439.

Ananthakrishnan, T. N. 1989 *Thrips and Gall dynamics.* Oxford and IBH, New Delhi, 120 pp.

Ananthakrishnan, T. N. 1990 *Reproductive biology of Thrips.* Oak Park, Michigan, 158 pp.

Ananthakrishnan, T. N. 2001 Phytochemical defense profiles in insect-plant interactions. *In.: Insect and Plant defense dynamics.* Ed. T. N. Ananthakrishnan, Oxford & IBH, New Delhi: 1-21.

Ananthasubramanian, K.S. and Ananthakrishnan, T. N. 1975 Taxonomic, Biological and Ecological studies of some Indian membracids (Insecta: Homoptera), pt. I. *Rec. Zool. Surv. India,* **68**: 161-272.

Baldwin, I. T. 1999 Induced nicotine in native *Nicotiana* as an example of adaptive phenotypic plasticity. *J. Chem. Ecol.,* **25**:3-30.

Becker, S. D. 1963 *Animal photoperiodism.* Holt, Reinhart and Winston, Inc., N. York: 71-75.

Bernays, E. A. and Chapman, R. F. 1987 Evolution of deterrent responses in plant feeding insects. In.: R. F. Chapman, E. A. Benrnays and J. E. Stoffolano (ed.) *Perspectives in* Chemoreceptors *and behaviour.* Springer Verlag, New York: 159-173.

Bernays, E.A and Chapman, R.F, Bradshaw, A.D. 1973 Environment and phenotypic plasticity. Brookhaven Symposia. **25**: 75-94.

Bowers, M. D. and Puttick, G. M. 1988 Response of specialist and generalist insects to quantitative allelochemical variations. *J. chem. Ecol.,* **14**: 319 - 324.

Bradshaw, A. D. 1965 Evolutionary significance of phenotypic plasticity in plants. Adv. Genet. **13**:115-155.

Casey, T. M. 1993 Effects of temperature in foraging caterpillar. In: *Caterpillar,* ed. Stamp, N. E. and Casey, T. M. Chapman and Hall, London: 5-28.

Chapman, R. F. and Lee, J. C. 1991 Environmental effects on members of peripheral chemoreceptors on the antennae of a grasshopper. *Chem. sciences.* **16**: 607-616.

Chauvin, R. 1967 *The world of an insect.* World University Library. McGraw-Hill Book Company. New York, 254pp.

Claridge, M. F and J. den Hollandu 1983 The biotype concept and its application to insect pests of Agriculture. *Crop Protection,* **2**:85-95.

Crespi, B.J. 1982 Territoriality and fighting in a colonial thrips *Hoplothrips pedicularius* and sexual dimorphism in Thysanoptera. *Ecol. Entomol.,* **11**: 119-130.

Crespi, B.J. 1988 Adaptation, compromise and constraint: Redevelopment, morphometrics and behavioral basis of fighter-flier polymorphism in male *Hoplothrips karnyi. Behavioral Ecology and sociobiology,* **23**; 93-104.

Crespi, B.J. 1990 Subsociality and female reproductive success in a mycophagous thrips: an observational and experimental analysis. *J. Insect Behav.* **3**: 61-74.

Crespi, B.J. 1992 Eusociality in Australian gall-thrips. Nature, **359**: 724-726.

Dale, J. F. and Tobes, S. S. 1990 The endocrine basis of locust phase polymorphism. Pp. 393 - 414 in R. Chapman and A. Joern (Eds.). Biology of Grasshoppers. Wiley, New York. 563 pp.

Dean, J. M. 1982 Control of diapause induction by a change in *Melanoplus sanguinipes*. *J. Insect Physiology* **28**: 1035-1040.

Dingle, H., Mousseau, T.A. and Scott, S. M. 1990 Altitudinal variation in life cycle syndromes of California populations of the grasshopper, *Melanoplus sanguinipes* (F). *Oecologia* **84**: 199-206.

Dobzhansky, T. 1971 *Genetics of evolutionary process.* Columbia University Press, New York. 505 pp.

Emlen, D. J. and Nijhout, H. F. 2000 The development and evolution of exaggerated morphologies in insects. Annual Review of Entomology 45: 661-708.

Fiedler, K. 1996 Host plant relationships of lycaenid butterflies: Large-scale patterns, interaction with plant chemicals and mutualism with ants. *Ent. exp. appl.,* 80:259-267.

Grant, A., Hassall, M. and Willott, S. 1993 An alternative theory of grasshopper life cycles. *Oikos* 66: 263 -268.

Greenfield, M. D., and Shelly, T.E. 1985 Alternative mating strategies in a desert grasshopper; evidence of density dependence. Animal Behaviour 33: 1192 -1210.

Greenfield, M. D., and Shelly, T.E. 1990 Territory-based mating systems in desert grasshoppers. Pp. 315-336 in R. Chapman and A. Jeorn (Eds.). Biology of Grasshoppers. Wiley, New York. 563 pp.

Geervliet, J. B. F., Ariem, S . Dicke, M. and Vet, L. E. M 1998 Long distance assessment of patch profitability high volatile infochemical by the parasitoid *Cotesia glomerata* and *C. reubecola* (Hymenoptera: Braconidae). *Biological control,*11: 113-121.

Hardie, J. and A. D. Lees 1985 Endocrine control of polymorphism and polyphenism. In: G. A. Kerkut and L. A. Gilbert (eds.) Comprehensive insect physiology, biochemistry and pharmacology. Vol.8.Pergamon, Oxord : 441-490.

Hosino, Y. 1940 Genetical studies on the pattern types of the lady bird beetle, *Harmonia axyridis* Pallas. J. Genetics, **40**: 215-228.

Johnson, K. L. 1982 Sexual selection in a brentid weevil. *Ent. Mitt.,* **12**: 225-232.

Kapur, A. P. 1959 Geographical variations in colour patterns of some Indian ladybird beetles (Coccinellidae: Coleoptera). I. *Coccinella septumpunctata* Linn., *C. transvarialis* and *Coleophora ocellata* Myers. *Proc. I. All India Congress, zool.* pt. 2:479-492.

Karban, R. and Myers 1989 Induced plant responses to herbivores. *Ann. Rev. Ecol. Syst.,* **20**:331-348.

Karban, R. and Baldwin, I. T. 1997 *Induced responses of herbivores.* The University of Chicago Press. Chicago, 319pp.

Lewis, W. J. and Sheehan, W. 1997 Parasitoid foraging from a multitrophic perspective: Significance for biological control. In: *Ecological Interaction and Biological Control* (ed.) D. A. Andrew, D. W. Rigsdale and R. F. Nyvall. Westview Press. Boulder: 277-281.

Lee, M. and Whitman, D.W. 2005. Age of sexual receptive and gonadal development in the lubber grasshopper, *Romalea micmptera. J. Orthoptera Research* 13 : (in Press)

Mayr, E. 1963 Animal species and their evolution. Harvard Univ. Press, Cambridge.

Meyers, L. A. and J. J. Bull 2002 Fighting change with change: adaptive variations in an uncertain world. *Trends in Ecology and Evolution,* **17**: 551-557.

Mound, L.A. 1970 Sex intergrades in Thysanoptera. *Entomol. Mon. Mag.,* **105**: 186-189.

Mound, L. A. 1971 The complex of Thysanoptera rolled leaf galls on *Geijera. J.Austal. Ent. Soc.,* **10**: 83-97.

Mound, L. A. 1977 Species diversity and the systematics of some New World leaf litter Thysanoptera. *Syst. Entomol.,* **2**: 225-244.

Mound, L. A. 1994 Gall-inducing Thysanoptera: a search for pattern - chapter 8 in M.A.J. Williams (ed.) *Plant galls: Organisms, interactions, populations.* Oxford University Press: 131-149.

Mound, L. A. 1995 Biosystematics of two gall-inducing thrips with soldiers (Insecta: Thysanoptera) from Acacia trees in Australia. *J. Nat. Hist., 29*: 147-157.

Mound, L. A. and Crespi, B. J. 1995 Biological studies of two gall-inducing thrips with soldiers defending the colony (Insecta: Thysanoptera), for Acacia trees in ambalia. *J. Nat. Hist., 29*: 147-157.

Mound, L. A., B. J. Crespi and Tucker, A. 1998 Polymorphism and kleptoparasitism in Thrips (Thysanoptera: Phlaeothripidae) from woody galls on Casuarina trees. *Aust. J. Ent.,* 37:8-16.

Mousseau, T.T. and Fox, C.W. 1998 Maternal Effects as Adaptations. Oxford Univ. Press, New York. 375 pp.

Nagaraja, H. and Sankaran, T. 1995 Genetic diversity of Trichogrammatidae and its implications in biological control. In: *Biological control of social forest and plantation crops insects.* Ed. T. N. Ananthakrishnan, Oxford & IBH, New Delhi: 179-191.

Nanjundiah, V. 2003 Phylogenetic plasticity and evolution of genetic assimilation. In: Origination of organismal form (ed.) E. B. Miiller and S. A. Newman, M. I. T. Press, 246-263.

Nylin, S. and Gotthard, K. 1998 Plasticity in life-history traits. *Annual Review of Entomology* **43:** 63-83.

Pigliucci, M 1996 How organisms respond to environmental changes: from phenotypes to molecules (and vice versa). *Trends in Evolution and Ecology.* **11:** 168-173.

Pigliucci, M. 2001 Phenotypic Plasticity: Beyond Nature and Nurture (Syntheses in Ecology and Evolution). John Hopkins. Pp. 384 pp.

Relyea, R.A. 2002 Costs of Phenotypic plasticity. *Amer. Nat.* **159:** 272-282.

Rhoades, D. F. 1985 Offensive and defensive interactions between herbivores and plants and their relevance in herbivore population dynamics and ecological theory. *Am. Nat.,* **125**:205-238.

Ritchi, M. G., Bulin, R. K. and Hewit, G. M. 1987 Causation, fitness effects and morphology of macropterism in *Chorthippus parallelus* (Orthoptera: Acrididae). *Ecological Entomology* **12:** 209-218.

Scheiner, S. M. 1993 Genetics and evolution of phenotypic plasticity. *Annual Review of Ecology and Systematics* **24**: 35 - 68.

Schlichting, C. D. and Pigliucci, M 1998 Phenotypic Evolution: A Reaction Norm and Perspective. Sinauer, Sunderland, MA. 340 pp.

Scriber, J. M. 1983 Evolution of feeding specialization, physiological efficiency and host races in selected Papilionidae and Saturniidae. *In*: R. F. Denno and M. S. McClure (ed.). *Variable Plants and Herbivores in natural and managed systems.* Academic Press, New York.

Simms, E. L. 1998 *Genetic structure and local adaptations in natural insect populations.* Chapman Hall, New York.

Smith, C. M. 2000 Plant resistance to insects. In: *Biological and bioecological control of insect pests.* eds. Rechcigl, J. E. and N. A. Rechcigl. Lewis Publication, New York:171-208.

Southwood, T. R.E. 1961. A hormonal theory of the mechanism of wing polymorphism in heteroptera. *Proc. R. Ent. Soc.,* London, **36A:** 63-70.

Stamp, N. E. 1993 A temperate region view of the interaction of temperature, food quality and predators of caterpillar foraging. In. *Caterpillars,* eds Stamp, N. E. and Casey T. M. Chapman Hall, London: 478-508.

Stamp, N. E. & Casey, T. M. 1993 *Caterpillars.* Chapman Hall, London, 587pp.

Stauffer, T. W. and Whitman, D.W. 1997 Grasshopper Oviposition. Pp. 231 - 280 in S. Gangwere, M. Muralirangan (Eds.). *The Bionomics of Grasshoppers, Katydids, and their Kin.* CAB International, Wallingford, UK. 529 pp.

Stearns, S. C. 1989 The evolutionary significance of phenotypic plasticity. *Biosource,* **39**: 436 - 445.

Tallamy, D. W. and M. J. Ranpp 1991 *Phytochemical induction by Herbivores.* John Wiley, New York.

Tatar, M. and C. M. Yin 2001 Slow aging during insect reproduction diapause: Why butterflies, grasshoppers and flies are like worms. *Exp. Gerontology,* **36**: 723 - 738.

Tauber, M. J., Tauber, C. A. and Masaki, S. 1896. Seasonal Adaptations. Oxford Univ. Press. New York. 411 pp.

Thompson, D. B. 1992 Consumption rates and the evolution of diet-indused plasticity in the head morphology of *Melnoplus femurrubrum* (Orthoptera: Acrididae). *Oecologia* **89**: 204 - 213

Thompson, J. N. 1994 *The co-evolutionary process.* The University of Chicago Press, Chicago. 376pp.

Uvarov, B. 1996 Grasshoppers and Locusts. Vol. I. Anti-Locust Research Centre, Cambridge. 481 pp.

Uvarov, B. 1996 Grasshoppers and Locusts. Vol. II. Centre for Overseas Pest Research, Cambridge. 613 pp.

Varadarasan, S. and Ananthakrishnan, T. N. 1981 Biological studies on some gall thrips. *Proc. Indian natn. Sci. Acad.,* B, **48**: 35 - 43.

Vet, L. E. M. & Dicke, M. 1992 Ecology of infochemical use by natural enemies in a tritrophic context. *Amer. Rev. Entomol.,* **37**: 141-172.

Via, S., Gomulkiewicz, R., E., Scheiner, S. M. Schhichting, C. D. and P.H. Van Tienderen 1995 Adaptive phenotype plasticity: De Jong, Consensus and controversy. *TREE,* **10**: 212-217.

Waddington, C. H. 1956 Genetic assimilation of the bithorax phenotype. *Evolution,* **10**: 1-13.

Wigglesworth, V. B. 1961 *Insect Polymorphism-* a tentative synthesis. Symposium No. 1, Royal; Entomological Society, London.

Wood, T.K. and Guttman 1981 The role of host plants in the speciation of tree hoppers: an example from the *Enchenopa binotata* complex. In: *Insect life history patterns.* R. F. Denno and H. Dingle. Springer Heidelberg:39-54.

West-Eberhard, M. J. 2003 Developmental Plasticity and Evolution. Oxford University Press, Oxford. 794 pp.

Whitman, D. W. 1998 Plant natural products as pasitoid cuing agents. Pp. 386-396 in H. C. Cutler (Ed.). Biologically Active Natural Products Potential use in Agriculture. ACS Symposium Series 380. American Chemical Society, Washington, DC.

Whitman, D. W. and Nordlund, D. A. 1994 Plant chemicals and the location of herbivorous arthropods by their natural enemies. Pp. 135-159 in T. N. Ananthakrishnan (Ed.). Functional Dynamics of Phytophagous Insects. Oxford & IBH, New Delhi. 304 pp.

2

Phenotypic Plasticity of Host Selection in Adult Tiger Swallowtail Butterflies, *Papilio glaucus* L. (Lepidoptera: Papilionidae)

Rodrigo J. Mercader and J. Mark Scriber
Department of Entomology, Michigan State University, East Lansing, MI 48824 USA, Scriber@msu.edu

Introduction

Phenotypic plasticity as defined by Pigliucci (2001) is, "the property of a given genotype to produce different phenotypes in response to distinct environmental conditions" implying a distinct predictable impact of certain environmental conditions rather than random noise. The non-genetic environmental induction of host plant preference (whether an increased 'specificity' or a change in 'rank order'; Courtney and Kibota 1990) is a classical example of phenotypic plasticity (Agrawal 2001). Phenotypic plasticity has been observed at least since the study of the genetic control of traits was begun in the early 20th century. However, the factors causing the differential expression of genotypes have remained a challenge in most systems. A key problem is the difficulty of separating the various environmental factors organisms encounter in nature from the genetic variability present in populations (Mousseau and Roff 1987, Mousseau et al. 2000). In some situations, environmentally induced (i.e. plastic) traits may even become fixed (genetically invariant) in populations via selection for stable expression of the trait under new conditions so that it no longer requires the original environmental stimulus (i.e. 'genetic assimilation'; Pigliucci and Murren 2003). Further complications arise when attempting to identify the environmental factors inducing the differential expression, and the mechanism(s) that facilitate this induction.

Many evaluations of phenotypic plasticity (or the capacity for it) have focused primarily upon developmental plasticity, with irreversible variation in traits expressed in one form or another during development due to particular kinds of environmental variation, and described by reaction norms (Stearns 1989, Via et al. 1995, Schlichting and Pigliucci 1998). This irreversible plasticity would include the concept of 'seasonal polyphenisms' in which different generations have discrete phenotypes during the season (e.g. Koch 1992, Shapiro 1976, Danks 1999, Beldade and Bakerfield 2002). In contrast to these 'irreversible' developmental changes, 'phenotypic flexibility' has been proposed to cover the reversible metabolic and/or endocrine switches within single organisms (Nation 2002, Wilson and Franklin 2002, Piersma and Drent 2003). An example of this concept is adult 'learning' in oviposition (i.e. Preference induction or aversion; Miller and Stricler 1984, Papaj and Prokopy 1989, Szentsi and Jermy 1990, Papaj and Lewis 1993, Cunningham et al. 1998). These forms of phenotypic plasticity can be highly adaptive under certain environmental conditions. For example, in a model by West and Cunningham (2002) learning can be favored in post-alighting behavior when a host limitation is present. The results from their model are consistent with findings from the generalist *Helicoverpa armigera* (Cunningham et al. 1998). Furthermore, Weiss and Papaj (2003) found that the pipevine swallowtail, *Battus philenor*, was able to simultaneously learn color cues for oviposition and nectaring resources. However, in the same study they found that training within one context biased what was learned in the other context, suggesting an information processing constraint.

The relative acceptability of host plants for adult oviposition by herbivorous insects is determined by a balance of numerous internal and external stimulants/deterrents (Miller and Strickler 1984, Bossart and Scriber 1999). External factors may include host plant volatiles, surface chemistry, color, texture, and shape (Rausher 1978, Brown et al. 1981, Miller and Strickler 1984, Harris and Rose 1990, Renwick 1990, Rewick and Chew 1993; Huang and Renwick 1995; Landolt and Molina 1996, Carter and Feeny 1999, Carter et al 1999, Frankfater and Scriber 1999, 2003). In addition, the choice of host plant (or plant part) by an ovipositing female may be influenced by the microclimate (humidity, temperature, light intensity, or other factors; Grossmueller and Lederhouse 1986, Scriber and Lederhouse 1992) including seasonal thermal constraints (Scriber 1996b, 2002b) as well as associated communities of microorganisms, parasites, predators, and competitors (Jermy 1988, Bernays and Graham 1988, Denno et al. 1995, Niemala and Mattson 1996, Ohsaki and Sato 1999, Redman and Scriber 2000), and availability of host plants (Rausher 1978, Kareiva 1985, Fitt 1986,

Scriber 1986).Recently, plants have been shown to exhibit induced defensive responses to insect egg deposition by females (via hypersensitive responses and/or release of oviposition deterrents; Hilker and Meiners 2002).

Internal motivational factors affecting host selection by ovipositing adults and motivation to oviposit (Miller and Strickler 1984, Lederhouse and Scriber 1987, Spencer et al. 1995) include: insect age (Rausher 1983), egg loads accumulated inside the female or on the host plant (Jones 1977, Fitt 1986, Minkenberg et al. 1992, Prokopy et al. 1994), length of time since last oviposition (Singer 1983), and time since last mating (Spencer et al. 1995). The extensive variability in external and internal factors affecting host selection makes the determination of locally ideal hosts difficult at best for researchers hoping to identify particular mechanisms underlying phenotypic plasticity (Papaj and Rausher 1987, Thompson 1988a, Thompson and Pellmyr 1991, Bernays 1991, Nylin et al 1996, Nylin and Gotthard 1998, Withers et al 1998, Scriber 1996b, 2002b). In recent years the evaluation of 'adaptive phenotypic plasticity' in evolutionary ecology has been a subject of much attention, often with debate regarding unclear concept definitions such as whether plasticity must always be adaptive (Via et al. 1995, Dewitt et al. 1998, Schlichting and Pigliucci 1998, Agrawal 2001, Piersma and Drent 2003).

Host selection by adult phytophagous insects with relatively sessile larvae that complete their development on the host selected by the mother, such as most swallowtail butterflies, is expected to be under high selective pressure. Therefore we might expect little within population phenotypic variability in traits associated with host selection. However, in species with large continuous ranges the geographic and temporal variation in host plant suitability is likely to require plastic responses by local populations whether through strong local adaptation or phenotypic plasticity. In generalist species, the problems in sorting out genetic and non-genetic influences is predicted to be greater due to the increased variation in plant species suitability, associated communities of natural enemies, competitors, and abiotic factors (Scriber 2002a). For over 20 years the Scriber lab group and others have studied the geographic variation in oviposition preference of the most polyphagous swallowtail butterfly *Papilio glaucus* (L.), primarily looking at genetic differentiation. In this short review we will concentrate on the phenotypic plasticity in host selection of adult *Papilio glaucus* using information derived from lab based oviposition trials of butterflies collected throughout the *P. glaucus* range.

The General System

P. glaucus is the most polyphagous Papilionid butterfly in the world feeding on over 18 plants in seven families (Scriber 1984a). A distinct ovipositional preference for tulip tree, *L. tulipifera*, in the Magnoliaceae predominates throughout most of its range. However, in some areas of the range preference for hop tree, *Ptelea trifoliata* (Rutaceae), can be equal or greater than that for *L. tulipifera*; Scriber et al. 2003). In some extreme cases preference for a different host may be present, such as in southern Florida where only *M. virginiana* (Magnoliaceae) is present, and an equal preference for *M. virginiana* and *L. tulipifera* is found (Bossart and Scriber 1995a, b). Other common hosts used include black cherry *Prunus serotina* (Rosaceae), white ash *Fraxinus americana* (Oleaceae), and to a lesser extent sassafras, *Sassafras allbidum*, and spicebush, *Lindera benzoin* (Lauraceae) (Scriber et al. 1975). No single host plant covers the entire range of *P. glaucus*, and the distribution of these host plants form mosaics of host availability including regions where only one host is present (e.g. *M. virginiana* in southern Florida; Scriber 1986). In addition, as the season progresses host quality declines at different rates and can potentially alter ovipositional choice. An extreme example of this variance in space and time occurs in 'cold pockets' where degree day discordance between small regions and neighboring localities causes a break in the timing of *P. canadensis* emergence and host plant growth. Butterflies in these localities become isolated and have been noted to have different ovipositional preference compared to those in surrounding areas as explained by the 'voltinism-suitability model' (Scriber 2002a).

Not surprisingly the larvae of this species have a wide range of detoxification abilities that permit them to survive on a variety of chemically diverse plants (Scriber 1988, Nitao 1995, Li et al 2002a, 2002b). The suitability for larvae of the host selected by adults may be determined by complex interactions with abiotic factors such as the constraints of seasonal total degree-day thermal unit accumulations at given localities (Scriber and Hainze 1987, Nylin 1988, Scriber and Lederhouse 1992, Scriber 1996a, 2002a). The following text is based upon multi-choice oviposition arenas (mostly three-choice or seven-choice) as described in Scriber (1993, 1994) where daily repeatability of behavioral preference patterns was basically consistent and the number of eggs are large enough that experimental procedures were appropriate (Raffa et al. 2002).

Genetic Basis of Preference Hierarchy

Several genetic controls of ovipositional behavior in *Papilio* spp. have been discerned fairly successfully. An interesting aspect of inheritance in *Papilio* spp. is the high degree of sex linkage associated. In the polyphagous *P. zelicaon* ovipositional hierarchy (preference ranking of hosts) is determined by loci on the X chromosome and some loci not on the X chromosome (Thompson 1988b, 1993, 1994). In addition, the ovipositional preference hierarchy of *P. zelicaon* appears to have an exceptionally high level of conservatism throughout its large geographic range (Wehling and Thompson 1997). This conservatism in preference hierarchy has been attributed to a gene or genes in the X chromosome, while the small differences in specificity (degree of fidelity to host hierarchy) have been attributed to loci not on the X chromosome (Wehling and Thompson 1997).

Scriber (1994) reports the ovipositional behavior of *P. glaucus* and *P. canadensis* reciprocal F1 hybrid daughters to follow that of the paternal species. The adult butterflies were assayed in three choice arenas consisting of *L. tulipifera*, *Populus tremuloides*, and *Prunus serotina*; the first host being the preferred host of *P. glaucus*, the second the preferred host of *P. canadensis*, and the third a mutual host. The results of that indicate a distinct sex linkage of ovipositional preferences much like in *P. zelicaon*, at least for the three hosts assayed. Given the wide range of phytochemical variety of the hosts utilized by *P. glaucus*, and the geographic variation in preference profiles, it is difficult to presume a single gene is involved in determining ovipositional hierarchy. However, if several genes are involved in determining ovipositional hierarchies the hybrid crosses indicate that these are likely to be closely linked and on the X chromosome. This X-linked control of the preference hierarchy has also been shown to hold for seven-choice arrays as well as three-choice arrays, with hybrid daughters exhibiting a preference profile that is nearly identical to their paternal (*P. glaucus*) species versus the pattern seen in the maternal *P. canadensis* species (Scriber, unpublished data). Sex linkage of ovipositional behavior may be a common occurrence in the Lepidoptera. For example, in the unrelated Nymphalid *Polygonia c. album* Janz (1998) found species sex linkage was involved in determining ovipositional behavior. In this situation differences in 'specificity' were attributed to a gene or genes on the X chromosome leading to differentiation between specialists and generalists insects.

Bossart and Scriber (1995b) found significant genetic variation in ovipositional specificity between populations from Ohio and Georgia compared to a Southern Florida population. The southern Florida population has *M. virginiana* as the sole host present in the region and it is towards this host that greater specificity is found in that population. Despite this difference extensive gene flow, based on allozyme frequencies, was found in the same study. The inferences from that study indicate that (based on the strong selective forces present and the lack of strong genetic barriers) a mosaic of genotypes is present throughout the *P. glaucus* range. There also appears to be significant variation in preference in *P. glaucus* due to genetic differentiation within populations (Scriber et al. 1991a, Bossart and Scriber 1995b).

Behavioural Induction of Adults

Change in preference by ovipositing adults due to previous experience has been documented in various insects (e.g. Vinson et al 1977, Rausher 1978, Prokopy et al. 1982, Mark 1982, Jaenike 1983, Rausher 1983, Stanton 1984, Cooley et al. 1986, Szentesi and Jermy 1990, Cunningham 1998, Bjorksten and Hoffmann 1998; see also Agrawal et al. 2002). In particular, the fairly well studied *B. philenor*, in the same family as *P. glaucus*, indicates that neurological constraints are an unlikely possibility in *P. glaucus*. Scriber (1993) exposed *P. glaucus* adults from the same population for two days to either *L. tulipifera*, *P. trifoliata*, *P. tremuloides*, or *P. serotina*. These butterflies along with unexposed butterflies were then presented with a four choice oviposition bioassay consisting of the four hosts mentioned above. The results from this data indicated a lack of behavioral induction of preference rank hierarchy. However, a reanalysis of the same data indicates a significant effect of exposure on the ovipositional specificity of these butterflies on the first day after being placed in the arena with the four hosts present (Fig. 1).

The reanalysis of the data consisted of a comparison of the number of eggs laid by butterflies in the different exposure groups the first day after they were placed in the choice arenas. The data was analyzed with host plant choice as the dependent variable, exposure as the explanatory variable, and total number of eggs laid per female as a covariate in a generalized linear model with a Poisson distribution and least squares as a means separation technique using Proc Genmod (SAS Institute). We ask for caution when interpreting patterns from these results, because the model fit indicated overdispersion may be an issue, and the sample size was fairly small. However, the analysis implies that adult experience influences the

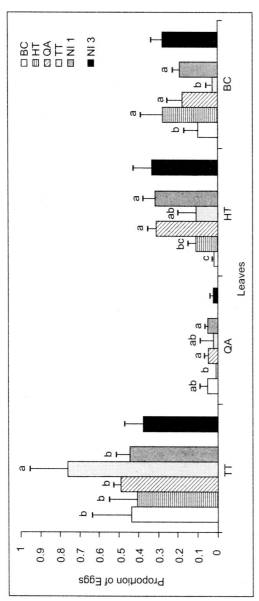

Fig. 1   Proportion of eggs laid the first day after being placed in oviposition arenas by *P. glaucus* butterflies exposed for two days with either black cherry, BC, (*Prunus serotina*); hop tree, HT, (*Ptelea trifoliata*); quaking aspen, QA, (*Populus tremuloides*); or tulip tree, TT, (*Liriodendron tulipifera*). NI 1 represents the response of *P. glaucus* butterflies the first day after being placed in the oviposition arena by butterflies not induced with any host. Analysis of the data was run as an ANOVA with poisson error using total number of eggs laid as a covariate and mean separation was performed by protected Fishers LSD within host. Proportions of eggs laid per host are presented for ease of visual interpretation; means with different letters are different at the 0.05 type I error level for the proportion of eggs laid on that host. NI 3 represents the response of *P. glaucus* butterflies the third day after being placed in the oviposition arena by butterflies not induced with any host; presented to illustrate a similar pattern present the following days by non-induced butterflies, an indication that deprivation does not explain the patterns observed.

specificity of host choice in *P. glaucus*. This result is particularly evident when unexposed butterflies and those exposed with *L. tulipifera* are compared (Fig. 1).

As the butterflies used in these assays were exposed for two days those exposed to non-hosts (*P. tremuloides*) or lesser preferred hosts may have experienced deprivation, which according to motivational models (Miller and Strickler 1984) tend to decrease specificity. In this study such a change was not observed (Fig. 1; compare NI 1 and NI 3), although it also cannot be entirely discounted due to the low replication. However, the greatest impact in specificity was an increase in specificity to *L. tulipifera* by individuals exposed to *L. tulipifera*.

Several theoretical reasons for why this phenomenon may be selectively favored have been proposed. Courtney and Kibota (1990) used ideas based on foraging theory for predators to explain preference induction. Within this framework changes in specificity should be linked directly to the availability of preferred hosts. West and Cunningham (2002) subdivided the concept of optimal foraging for adults by building a simple two-host model based on both egg load limitation and host limitation. In this situation egg load limitation and host limitation act as counterbalancing forces, egg load limitation pushing towards 'genetically programmed' specificity and host limitation towards a more plastic specificity. In *P. glaucus*, where eggs are laid singly and females often lay over 100 eggs, host limitation is more likely to be a selective force than egg load limitation. The results presented in Fig. 1 and the mosaic distribution of host plants are consistent with the notion that preferred host scarcity/limitation may be selecting for the maintenance of plasticity in adult host selection.

Behavioural Induction at Larval Stage

The induction of larval preference/deterrence has been documented in various insects (e.g. Bernays and Weiss 1996, Renwick and Lopez 1999) and is fairly well accepted; however the same cannot be said for the induction of adult preference/deterrence during the larval stage. A classic theoretical mechanism of phenotypic plasticity is the Hopkins Host Selection Principle which proposes that the ovipositional behavior of adults is influenced by the larval environment. Although Hopkins, original definition and the early tests of this principle did not separate between behavioral induction at the larval stage and local adaptation and/or sympatric speciation (Reviewed in Barron 2001), the idea of behavioral induction of adults during their larval stage has been what is directly associated with Hopkins Host Selection

Principle. This concept has had a fair amount of support, but its association with pre-imaginal conditioning has led to a general rejection of the term (Reviewed in Barron 2001). In particular, the issue of the memory retention after the extensive reorganization of the insect brain during metamorphosis in holometabolous insects has brought upon a significant amount of criticism. However, Ray (1999) working with house flies exposed as larvae to sawdust scented with peppermint oil or geraniol was able to transfer the conditioned response of exposed adult flies to unexposed control adult flies by grafting neural tissue from adult exposed flies to adult unexposed flies. These results are a strong indication that memory may be retained after metamorphosis in holometabolous insects through cell survival. The prevalence and strength of this phenomenon in nature is unknown, but its potential impact upon local adaptation and speciation maintain this concept as intriguing at the very least.

A particular difficulty in identifying larval conditioning is separating early adult experience from larval experience. This difficulty led to a period in which the presence of larval induction was largely discounted (e.g. Courtney and Kibota 1990, Szentesi and Jermy 1990, van Emden et al. 1996, Barron 2001). However, several studies in which larvae have been exposed to specific host plant chemicals have indicated that induction of adult ovipositional preference may occur during the larval period (e.g. Del Campo et al. 2001, Akhtar and Isman 2003; see also Jaenike 1983, Tully 1994).

Bossart and Scriber (1999) reared *P. glaucus* on *L. tulipifera*, *M. virginiana*, or *P. serotina* and the preference of the resulting adults was assayed in two experiments. One experiment compared *L. tulipifera* reared butterflies and *M. virginiana* reared butterflies in two choice assays of those hosts, and the second experiment compared *L. tulipifera* reared butterflies and *P. serotina* reared butterflies in two choice assays of those two hosts. These results indicated no significant differences between *M. virginiana* and *L. tulipifera* reared butterflies. However, a significant difference in specificity between *P. serotina* and *L. tulipifera* reared butterflies was found. The disparity between the two experiments raises several questions regarding the factors required to induce preferences. These results could potentially be caused by plant chemistry differences and physiological constraints, or nutritional differences affecting egg production (discussed below). Both *M. virginiana* and *L. tulipifera* are in the Magnoliaceae and are likely to have similar secondary chemistry, while *P. serotina* is in the Rosaceae and likely to have fairly different chemistry. The potential chemical similarity between *L. tulipifera* and *M. virginiana* may preclude the ability to distinguish between two chemically related hosts based on an induction of preference, even if an induction is present. To be able to test such a hypothesis, a finer detail of the physiological basis of induction in larvae than currently exists is required.

These studies do not rule out the possibility of chemical traces in the hemolymph or pupal casing eliciting behavioral induction in the adult stage as described by the chemical legacy hypothesis (Corbet 1985). However, the ecological relevance remains identical regardless of the mechanism involved, as the shift in preference by the ovipositing adult would be based upon the host the larvae developed on. In addition to larval performance the importance of female performance based on the host it developed upon need to be considered as factors involved in ovipositional preference. For example, adult female longevity and fecundity in the agromyzid fly *Chromatomyia nigra* have been closely correlated with ovipositional hierarchy (Scheirs et al. 2000).

Fecundity

Specificity of ovipositional preference would be expected to be correlated with fecundity, because the cost of investment is inversely related to the number of eggs or egg batches that can be laid. For example, adult female longevity and fecundity in *Chromatomyia nigra* have been closely correlated with ovipositional hierarchy (Scheirs et al 2000). Various studies indicate a decrease in specificity with increasing egg load (e.g. Fitt 1986, Courtney et al. 1989, Odendal and Rausher 1990, Horton and Krysan 1991, Minkenberg et al. 1992, Jallow and Zalucki 1998,). Research in parasitoid insects has looked extensively into limiting factors affecting ovipositional behavior. This literature on parasitoid insects has traditionally separated parasitoid insects based on whether they were egg limited or time limited (now considered a shifting range rather than a dichotomy; e.g. Minkenberg et al. 1992, Rosenheim 1996, Heimpel & Rosenheim 1998). Egg limited parasitoids are expected to show high specificity due to offspring survival rate defining their fitness, while time limited parasitoids are defined by the optimal foraging theory. It is useful to consider this shifting dichotomy when considering phytophagous insects that complete their development within a single host (e.g. West and Cunningham 2002). The role of limitation by eggs or time to locate hosts can have significant implications towards the role of phenotypic plasticity. If eggs are a limiting resource the cost of accepting a lower ranking host would be significant, while if host availability is the main limiting resource the cost of not accepting a host would be greater.

In *P. glaucus* despite careful searching, no strong evidence has been found relating ovipositional preference and fecundity measured as number of eggs laid or size has been found (Bossart and Scriber 1999). Regressing *P. glaucus* ovipositional choice assays for *L. tulipifera*, *P. serotina*, and *P. tremuloides* for 159 butterflies collected between 2000 and 2003 indicates a significant negative correlation between the number of eggs laid and the proportion of eggs laid on the preferred host *L. tulipifera* (P= 0.0103). However, this relation explains very little of the variance in host choice observed (Adj. r^2 = 0.0412). Fig. 2 illustrates the remarkably weak relationship between the proportion of eggs laid on *L. tulipifera* and the total number of eggs laid.

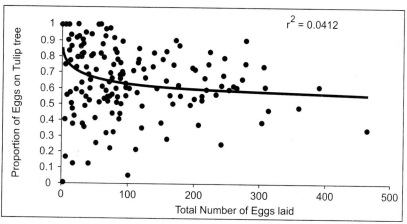

Fig. 2 Regression (r^2 = 0.04)of the proportion of eggs laid on tulip tree (*Liriodendron tulipifera*) and the total number of eggs laid by that butterfly in three-choice oviposition bioassays containing *L. tulipifera*, *Prunus serotina*, and *Populus tremuloides*. Females used in this regression were 159 butterflies collected from 12 different populations between 2000 and 2003. Populations were collected from the US states of Michigan, Georgia, Pennsylvania, Florida, and Virginia; all of which showed population preferences for *L. tulipifera*.

Papaj (2000) in a review of ovarian dynamics of insects points out that insects have a remarkable ability to mature eggs in an exceptionally fast manner, and with a high degree of control. This ability of insects may allow them to have decreased egg limitation since eggs can be fertilized as necessary with little waste. However, Rosenheim (2000) found that even in synovigenic insects (able to mature eggs throughout the adult stage), egg limitation can incur significant costs due to transient egg limitation, implying that egg limitation can potentially be a significant limitation under certain circumstances. Local environmental conditions (including latitudinally correlated factors) or maternal effects (Groeters and Dingle 1988, Ayers and Scriber 1994, Braby 1994, Parry et al. 2001) may significantly adjust egg size. In this sense, the nutritional quality of the parental hosts may have a significant impact upon potential egg loads or mating success. Other physiological changes, such as those associated with age, may also significantly affect the intensity of egg limitation.

Age

Age is expected to alter ovipositional specificity because it affects both the time available for host selection and the relative egg availability. *P. glaucus* adults collected from various populations were scored for wing wear class as a proxy for age and assayed on ovipositional arenas containing seven hosts. These assays indicated no significant effect of age class (Fig. 3). To reduce variability due to the presence of two preferred hosts in the assays (*L. tulipifera* and *P. trifoliata*), an analysis was run comparing the joint preference for the two hosts between age classes; again no significant difference among age classes was found ($F_{4,205} = 0.89$ P-value = 0.47).

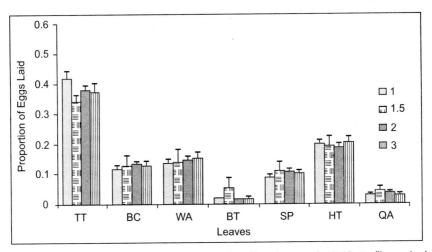

Fig. 3   Proportion of eggs laid in seven-choice oviposition bioassays by 210 butterflies ranked by age class collected in 1995 in the US states of Georgia, Ohio, and North Carolina. Age classes were determined by wing wear as in Scriber (1993). The seven hosts present in the bioassay were tulip tree TT (*Liriodendron tulipifera*), black cherry BC (*Prunus serotina*), white ash WA (*Fraxinus americana*), buckthorn BT (*Rhamnus frangula*), spice bush SP (*Lindera benzoin*), hop tree HT (*Ptelea trifoliata*), and quaking aspen QA (*Populus tremuloides*).

However, using lower number of choices, Scriber (1993) and Bossart and Scriber (1999) have both found some evidence of greater specificity with age; although the results in Bossart and Scriber (1999) were not statistically significant due to low power. In both of these studies the results appeared to indicate an increased specificity with age. The discrepancy between these studies and the results from the seven choice assays may be an indication of constraints in decision making due to large choice number (Bernays 2001), information value (Stephens 1989), or simply an artifact of the assays as the ability to detect differences rapidly declines with the number of choices presented to the study organism (Raffa et al. 2002).

Motivational state models (e.g. Miller and Stickler 1984) based on internal physiological changes predict that factors linked to time limitation (e.g. time since last oviposition) should increase the selection of suboptimal resources. However, in *P. glaucus* the apparent direction of age effects are towards a greater specificity. A reversal from time limitation to egg limitation appears counterintuitive as age increases, unless increased costs of egg production associated with aging. Rosenheim (1999) in an optimization model comparing egg and time limitations, using the probability of encountering better hosts as fitness currency, found the relative importance of egg mediated effects to increase with age.

'Mistakes'

Ovipositing females of swallowtail butterflies that place eggs on plant species that are toxic have been observed and called 'mistakes' (Straatman 1962, Wiklund 1975, Chew 1977, Berenbaum 1981, Larsson and Ekbom 1995, Renwick 2002, Graves and Shapiro 2003), but such events are believed to be extremely rare, especially for phytochemically-specialized species. Generalist such as *P. glaucus* are known to sometimes place a small fraction of their eggs on toxic hosts in nature (Brower 1958, 1959) and in controlled environments (Scriber et al. 1991b, Scriber 1993). In three choice oviposition assays, described above, with *L. tulipifera*, *P. serotina*, and *P. tremuloides* leaves, *P. glaucus* consistently makes two types of 'mistakes', laying eggs on portions of the arena containing no leaves at all or laying eggs on non-hosts.

Laying eggs on portions of the ovipositional arena containing no leaves may be simply due to an artifact of the ovipositional arena or a true reflection of a behavior inherent in *P. glaucus*. A residual level of excitation induced by host plant leaves present in the arena could produce the mistakes observed. If this scenario were to be correct, these mistakes would be an experimental artifact that should be correlated to a decrease in specificity. According to motivational models a decrease in specificity would predict an increase in mistakes, as a lower threshold between excitement and ovipositional bout would be expected (e.g. Miller and Stickler 1984). Alternatively if the number of these mistakes made is independent of specificity, there should be no significant relationship between specificity and the proportion of mistakes laid. Regressing the proportion of eggs laid on non-leaf portions of ovipositional arenas and the proportion of eggs laid on *L. tulipifera* are highly correlated (P-value < 0.001, Adj. $r^2 = 0.362$; Fig. 4). This is an indication that an overall reduction in specificity is the most likely cause for an increase in the number of mistakes made.

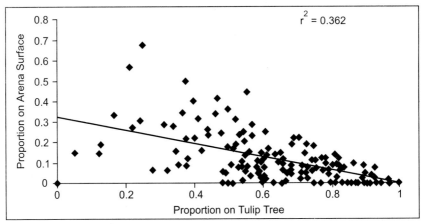

Fig. 4 Regression ($r^2 = 0.36$) of the proportion of eggs laid on tulip tree (*Liriodendron tulipifera*) and the proportion of eggs laid by that butterfly on non-leaf portions of the arena in three choice oviposition bioassays containing, *L. tulipifera*, *Prunus serotina*, and *Populus tremuloides*. Females used: see Figure 2.

As with the number of eggs laid on non-leaf areas, regressions between the proportion of eggs laid on other leaves would be expected to be highly negatively correlated to the number of eggs laid on other portions of the arena. For example, the proportion of eggs laid on *P. serotina* is highly correlated to those laid on *L. tulipifera* (P-value < 0.001, Adj. $r^2 = 0.603$). However this relationship is not linearly divided amongst leaves, as the correlation between the proportion of eggs laid on the non-host *P. tremuloides* and those laid on *L. tulipifera* in three choice tests was not nearly as strong (P-value <0.001, Adj. $r^2 = 0.088$; Fig 5). In fact, the proportion of eggs laid on portions of the oviposition arena with no leaves present was over twice that laid on *P. tremuloides* (Figs 4 & 5). This suggests that the previously described

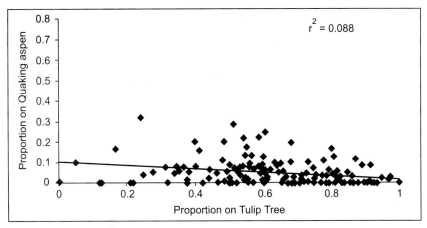

Fig. 5 Regression ($r^2 = 0.09$) of the proportion of eggs laid on tulip tree (*Liriodendron tulipifera*) and the proportion of eggs laid by that butterfly on quaking aspen (*Populus tremuloides*) in three choice oviposition bioassays containing, *L. tulipifera*, *Prunus serotina*, and *P. tremuloides*. Females used: see Figure 2.

deterrent effect of *P. tremuloides* (Scriber et al 1991a) may be independent of changes in specificity. This result has significant implications to motivational models as an overall decrease in specificity would predict a correlated increase in the number of eggs laid on *P. tremuloides*.

Comparing the relationship between *L. tulipifera* specificity and the proportion of eggs laid on non-leaf portions of the arena and the weak relationship with *P. tremuloides* indicates a qualitative difference in the type of mistakes made. This suggests that the number of eggs laid on non-hosts in nature may be independent of those separated amongst hosts.

The relationship between the proportion of eggs laid on *L. tulipifera* and those laid on *P. tremuloides* is even more striking when considering seven choice oviposition assays. Seven choice oviposition assays consisting of the hosts *L. tulipifera*, *F. americana*, *P. trifoliata*, *P. serotina*, *L. benzoin*, and the non-hosts *Rhamnus frangula* and *P. tremuloides* were performed for 211 adults in 1995. In this case the proportion of eggs laid on the preferred hosts *L. tulipifera* and *P. trifoliata* are significantly correlated to the proportion of eggs laid on lower ranking hosts and non-hosts (P-value <0.001 for all). However, as with three choice tests the variance explained is not evenly distributed amongst hosts (Table 1). In this case the number of eggs laid on non-leaf material was not counted and therefore it is not possible to determine whether a correlation between the number of mistakes and reductions in specificity are present. However, the lack of significant correlation amongst most hosts not including *L. tulipifera* and *P. trifoliata* (see Table 2) is an

Table 1   Data presented here is derived from seven-choice oviposition bioassays by 210 butterflies collected in 1995 in the US states of Georgia, Ohio, and North Carolina. The r^2 between tulip tree TT (*Liriodendron tulipifera*) and hop tree HT (*Ptelea trifoliata*), tulip tree TT, or hop tree HT and the remaining hosts are presented. The mean proportion of eggs laid on each of the hosts in the first column is displayed in the last column.

Host	r^2 TT and HT	r^2 TT	r^2 HT	Mean Proportion of Eggs laid
F. americana	0.37	0.28	0.003	0.144 ± 0.008
P. serotina	0.33	0.19	0.03	0.125 ± 0.007
L. benzoin	0.20	0.09	0.036	0.099 ± 0.006
P. tremuloides	0.17	0.11	0.004	0.032 ± 0.003
R. frangula	0.02	0.02	−0.003	0.017 ± 0.002

Table 2   P-values of pairwise Pearson correlations between the proportion of eggs laid on seven different hosts are presented. Data is derived from seven choice oviposition bioassays by 210 butterflies collected in 1995 in the US states of Georgia, Ohio, and North Carolina.

	L. tulipifera	P. trifoliata	F. americana	P. serotina	L. benzoin	P. tremuloides	R. frangula
L. tulipifera	1	<.0001	<.0001	<.0001	<.0001	<.0001	0.0151
P. trifoliata	X	1	0.1905	0.0066	0.0031	0.1671	0.5440
F. americana	X	X	1	0.7914	0.4821	0.1131	0.130
P. serotina	X	X	X	1	0.6415	0.3876	0.6598
L. benzoin	X	X	X	X	1	0.0649	0.8148
P. tremuloides	X	X	X	X	X	1	0.0009
R. frangula	X	X	X	X	X	X	1

indication that relationships amongst sub-optimal hosts were fairly independent. However, a strong correlation was found between the two non-hosts *R. frangula* and *P. tremuloides*, potentially an indication of a similarity in reaction to deterrents in non-hosts.

Conclusions/Discussion

The first general pattern that emerges from the studies reviewed is that host preference 'ranking' hierarchies are variable within populations yet generally appear to be genetically fixed in *P. glaucus* as a species. Unlike host rank hierarchy the degree of 'specificity' appears to be highly plastic as indicated by adult induction, larval induction, and age significantly impacting specificity. Physiological factors altering ovipositional behavior have been reviewed by various authors as motivational models (Miller and Strickler 1984, Spencer et al. 1995). An apparent general trend detected in these reviews is that insects have internal and external motivational inputs that moderate the acceptance of hosts. Recently Monks and Kelly (2003) reevaluated motivational models based on the lack of fit with their findings of learning in the diamondback moth. These authors divide responsiveness and acceptance into a two-stage incremental model, where the likelihood to accept a particular host (responsiveness) and the balance of non-specific cues for a host (acceptance) are separate steps. Within this incremental acceptance model sensitization to specific cues of one host in conjunction with an excitatory (internal) threshold can explain the induction of learned behavior. The results from the studies on *P. glaucus* adult and larval induction appear to support this model. However, the studies presented here were not designed to specifically test the incremental acceptance model and only offer peripheral support for it.

Within this model 'age' would appear as a non-specific inhibitory response, yet it appears to affect the specificity of ovipositional behavior. This difference does not conform to Monks and Kelly's model if it is to be taken at a general ovipositional behavior level. In addition, the correlation between eggs laid on non-leaf material and *L. tulipifera* in the three-choice tests described above (Mistakes) does not conform with Monks and Kelly's model. The number of eggs laid on non-leaf material appears to follow a reduction in overall specificity as would be predicted by motivational models. However, the lack of a strong correlation between the number of eggs laid on *P. tremuloides* and those laid on *L. tulipifera* does conform to Monks and Kelly's model and not to motivational models. The need for a general physiological model capable of accounting for stochastic as well as directional changes in ovipositional preference would greatly benefit attempts at understanding the adaptive value of phenotypic plasticity in host selection.

An interesting result is that fecundity (Fig. 2) does not appear to significantly affect ovipositional specificity in three-choice assays, but age did (Scriber 1993, Bossart and Scriber 1999). This dichotomy is most likely due to fecundity in this review being lifetime fecundity and not an age-specific fecundity function. The lack of a relationship between specificity and fecundity as mentioned earlier is an indication that, over the female's lifetime, host limitation is more likely to be a limiting factor than egg limitation.

Host limitation as modeled for time limited parasitoids can be an intense selection pressure even leading to selection for lower egg loads in favor of energy (Ellers and Jervis 2003). Although host plants tend to be a far less ephemeral resource than parasitoid hosts, the foliage of terrestrial autotrophs has very high carbon to nitrogen ratio, three times higher than freshwater communities (Elser et al. 2000, Sterner and Elser 2002), an indication of very low nutritional value. In addition, the highly diverse phytochemistry present in terrestrial systems (Schoonhoven et al. 1998) greatly reduces the value of the vast majority of plants to any single individual. Even in organisms long believed to be primarily driven by carbon to nutrient ratios such as *Daphnia*, recent results indicate that digestibility (of algae) had a stronger correlation than did carbon to phosphorous (main limiting resource) ratios in freshwater lakes (DeMott and Tessier 2002). A very high premium on locating valuable host resources should therefore be present since time to development and size have been highly correlated to important fitness components such as survival, mating success, egg load, and female longevity (Lederhouse et al. 1982, Wiklund et al. 1991, Ayres and Scriber 1994, Roff 1996, 2002, Zonneveld 1996, Nylin and Gotthard 1998).

Most phytophagous insects tend to specialize on particular tissues, often young tissues (Feeny 1970, Scriber and Slansky 1981, Slansky and Scriber 1985, Mattson and Scriber 1987) and as such even long lived plants act as ephemeral resources. For example, differences in host quality due to plant phenology have shifted local ovipositional hierarchies in *P. glaucus* towards hosts that are of otherwise lesser quality (Scriber 2002a). Despite the importance implied by host limitation as a selective factor and the high level of within population variation we fail to see heavy extents of local adaptation in *P. glaucus* except under extreme situations (e.g. southern Florida, cold pockets, and hybrid zones). This phenomenon could be due to the high gene flow amongst populations and variable temporal and spatial host resource mosaics, diffuse co-evolution, temporal genotype flooding, or a variety of scenarios. However, a complete discussion of population genetic variation is outside the scope of this short review (see Via 1990, 2001, Jaenike and Papaj 1992, Thompson 1994, 1998, Feeny 1995, Radtke and Singer 1995, deJong and Nielsen 2002, Scriber 2002a, 2002b, Via and Hawthorn 2002, Bossart 2003, Singer 2003).

What is within this scope is that, given the maintenance of a population level rank-order hierarchy, the presence of phenotypic plasticity in specificity has definite adaptive significance. The high levels of host variability encountered by a single genotype through multiple generations across varying habitat types would favor a flexible response to host selection. Through multiple generations in one season the quality of hosts will decline non-uniformly presenting a non-directional selective pressure. However, host plants that are of particularly high quality could create a surge in abundance of that insect genotype during the point in time when they are the best host available, skewing selection towards genotypes possessing a preference for that host and flexibility when the host is unavailable.

As presented by Rosenheim (1999) the fitness currency of accepting or rejecting a host can be considered to be the probability of encountering better hosts. The probability of encountering better hosts is directly linked to environmental predictability. It is difficult to presume a directional selective force would be present for an unpredictable resource. Unlike host availability (both temporal and spatial), host quality is a more predictable resource expressed in larval growth rates, developmental time, successful development to adult, and mating success. Bossart (2003) found significantly strong adult ovipositional preference and larval performance correlations in *P. glaucus* populations from Ohio where mixed groups of hosts including the preferred *L. tulipifera* are present. In contrast, in the same study a southern Florida population where only one host is present (*M. virginiana*; see Scriber 1986) a negative correlation between adult oviposition choice and larval performance was found. Here a predictable environment was present as only one host was present and a significant difference in host choice deviating from the highest quality host was found.

At an ecological level, the changes in specificity found could help explain the fairly homogeneous ovipositional rank-order hierarchies between *P. glaucus* populations (Fig 6), despite high levels of within population genetic variability (Fig 7; see also Scriber et al. 1991a, Bossart and Scriber 1995b). Acceptability, response to internal factors (threshold determination), and the rate of decay of the acceptability threshold are all likely to be the result of selection. Thus, within a single population, individuals that are generalist in their phenotypes may all have different genotypes for host-ranking and sppecificity determination (Courtney and Kibota 1990). Even in extreme cases where only one host is locally available, preference for *L. tulipifera* is maintained in populations (as exemplified by southern Florida populations which exhibit a close to normal distribution for ovipositional preference of *L. tulipifera* ; Bossart and Scriber 1995). Flexibility in specificity could weaken ovipositional thresholds or increase responsiveness to specific hosts to levels at which acceptance of available hosts is equal to that of the preferred host. However, a decrease in specificity may increase the overall level of mistakes made and increase the time required per ovipositional bout while experience is being gained and internal inhibitory controls reduced. This

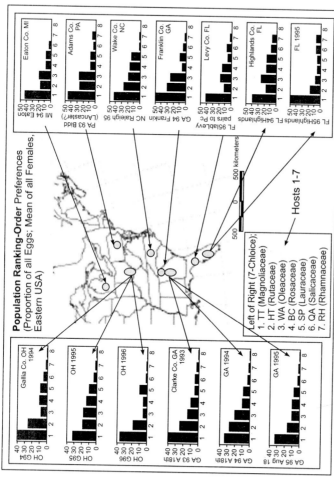

Fig. 6 The mean oviposition preference hierarchies of *Papilio glaucus* females for various populations across its geographic range; 1994 Eaton Co. Michigan (n=7 females), 1993 Adams Co. Pennsylvania (n=12 females), Gallia Co. Ohio (1994, n=17; 1995, n=46; 1996, n=15 females; all from August flights), 1995 Wake Co. North Carolina (n=17), 1994 Franklin Co. Georgia (n=7), Clarke Co. Georgia (1993, n=52; 1994, n=38; 1995, n=69 females), Levy Co. Florida (n=20), and Highlands Co. Florida (1994, top, n=6; 1995, bottom, n=3 females). Values for each plant species (tulip tree, hop tree, white ash, black cherry, spicebush, quaking aspen, *Rhamnus*, from left to right, or # 1 to # 7; 8=eggs placed on arena) are the grand means of egg proportions for all individual females (that laid 20 eggs or more).

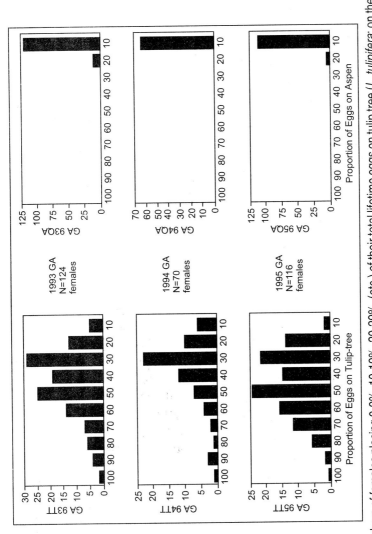

Fig. 7 The numbers of females placing 0-9%, 10-19%, 20-29%, (etc.) of their total lifetime eggs on tulip tree (*L. tulipifera*; on the left side) or on quaking aspen (*Populus tremuloides*; on the right side) in 7-choice oviposition arenas. The top, middle, and bottom figures represent females from Clarke Co. Georgia (1993, n=124; 1994, n=70; 1995, n=116 females). Females are all from August flights. Note the tremendous variability in strength of responses to tulip tree in contrast to the consistent (and near-total) avoidance responses to aspen.

potentially high fitness cost would require high gene flow and/or diffuse selection pressures to counter local adaptation.

The high degree of environmental variability altering relative host quality (e.g. phenology, chemistry, natural enemies) and availability (spatial and temporal) may reduce the directional strength of selection. Stephens (1989) applied economic models on the value of information (based on their variance) to optimal foraging models. A similar approach to the value of environmental 'information' on the selective pressures faced by genotypes may be useful in understanding the adaptive value of phenotypic plasticity. Fixed responses, directed plastic responses, and relatively stochastic responses may all be adaptive depending on environmental predictability. The ability to utilize environmental cues in host location (e.g. behavioral induction of adults and larvae) is likely adaptive only if the value of information (environmental predictability) is high. However, if the value of information is low, a fixed response (lack of plasticity) is more likely to be favored. Models based on information value and constraints may therefore prove to be useful in developing null expectations for the adaptive value of phenotypic plasticity.

Summary

In summary, oviposition preferences of insects are determined by a large array of external and internal stimuli. Natural selection has acted on the genetic basis of oviposition ranking behavior, but phenotypic plasticity also plays a very important functional role. Behavioral preference induction in adults and larvae, as well as the effects of fecundity, age, and other factors may modify the 'acceptability threshold' for different hosts and determine how far down a ranking-order the insect will go (i.e. its 'specificity').

The eastern tiger swallowtail butterfly (*P. glaucus*) is the most polyphagous of all 560 species of Papilionidae (Scriber 1984a; 1995). Here we show that the ranking order of oviposition preference for this generalist appears basically constant in seven family host choice arrays for populations from the northernmost to southernmost parts of its geographic range (Fig 6). Local 'phenotypic differentiation' (as opposed to local 'adaptation') may be accomplished largely by phenotypically flexible (plastic) 'specificity' in host selection by individual females, thereby alleviating the directional selection for rank-order changes that might be expected in various local populations. The lack of local adaptations for changing rank-order of preference is not due to a lack of genetic variation in oviposition behavior; it is clear that individual females vary in their placement, with 100% of their eggs to less than 10% of their eggs on (the putatively ancestral) tulip tree favorite (Fig. 7). 'Plasticity' appears to dampen (lessen the need for) directional selection (that might take a ranking-order change to fixation). Ovipositional flexibility for local hosts or ecological conditions is thus (theoretically) possible without need for extensive gene flow because of the specificity adjustments in this *Papilio* species. In contrast, the avoidance of aspen (both ranking order and specificity) may be genetically and behaviorally invariant, with almost every individual female placing few, if any eggs on this species (Fig 7).

The benefits of being polyphagous should exceed those of being specialized. The behavioral efficiency of locating the best host is generally less for generalists (perhaps due to neural limitations; Bernays 2001). Plasticity in behavior would be useful (e.g. where choices of hosts vary in availability in non-predictable fashion, as seen in generalists). Such non-genetic (inducible) responses may have permitted an ancient generalist (*P. glaucus*) to persist in ecologically and phytochemically diverse habitats rather than evolving numerous local host-specializations which would seem clearly possible, given the high levels of genetic variability for oviposition preferences that persists across the range of this generalist species (Bossart 2003, Bossart and Scriber 1995, 1999,; Fig. 7). Clearly, the complete ecology of the species needs to be addressed in order to understand the factors preserving this persistent polyphagy.

In addition to host plant and enemy-related factors, genetically-based behavioral oviposition divergence in time and location have been documented due to abiotic factors such as thermal selection as shown in highland/lowland *Drosophila* (Dahlgaard et al. 2001) and in climatic 'cold pockets' for *Papilio canadensis* (Scriber 2002b). As seen here with *P. glaucus*, it may be that plasticity in specificity allows different genetically-determined rank orders among related *Papilio* species (Scriber et al. 2003), but usually not within *Papilio* species. For example, heritable variation has been shown to exist for *P. glaucus* toward less preferred hosts rather than ranking order (Bossart and Scriber 1999). Both ranking and specificity of butterfly oviposition may be largely X-linked and genetically-controlled (Janz 2003; Singer 2003). However, the role of phenotypic plasticity in determining oviposition variation observed between individuals, populations, and congeneric species will continue to be an important prerequisite for our understanding of host plant (and adult preference/larval performance correlations; Mayhew 2001; Bossart 2003) relationships in phtyophagous insects.

Acknowledgements

We thank the following people for their help in the lab and/or field; Mark Deering, Jessica Deering, Jen Donovan, Piera Giroux, Holly Hereau, Matt Lehnert, James Maudsley, Jenny Muehlhaus, Michelle Oberlin, Gabe Ording, Laura Palombi, and Aram Stump. Research on Papilionidae has been supported in part by the Michigan Agricultural Experiment Station (Project MICL01644) and the National Science Foundation (LTER BSR-8702332; Ecology DEB-92201122; Ecosystems DEB-9510044; Population Biology DEB-9981608).

References

Agrawal, A.A. 2001. Phenotypic plasticity in the interactions and evolution of species. Science 294:321- 326.

Agrawal, A.A., J.K. Conner, M.T. J. Johnson, and R. Wallsgrove. 2002. Ecological genetics of an induced plant defense against herbivores: Additive genetic variance and costs of phenotypic plasticity. Evolution 56: 2206-2213.

Agrawal, A. A., F. Vala, and M.W Sabelis. 2002. Induction of preference and performance after acclimation to novel hosts in a phytophagous spider mite: adaptive plasticity? Am. Nat. 59: 553- 565

Akhtar, Y. and M. B. Isman. 2003. Larval exposure to oviposition deterrents alters subsequent oviposition behavior in generalist, *Trichoplusia ni* and specialist *Plutella xylostella* moths. J. Chem. Ecol. 29 (8): 1853-1870

Arendt, J. Adaptive intrinsic growth rates: an integration across taxa. Q. Rev. Bio. 72: 149-177.

Ayres, M.P. and J.M. Scriber. 1994. Local adaptations to regional climates in *Papilio canadensis* (Lepidoptera:Papilionidae). Ecol. Monogr. 64: 465-482.

Ayres, M.P., J. L.Bossart, and J.M. Scriber. 1991. Variation in the nutritional physiology of tree feeding swallowtail caterpillars. pp. 85-102. *In* Y.N. Baranchikov, W.J. Mattson, F.P. Hain and T.L. Payne [eds.]. Forest Insect Guilds: Patterns of Interactions with Host Trees. IUFRO Conference Proceedings (August 12-20, 1989) Abakan, Siberia, U.S.S.R. (USDA Forest Service Gen. Tech. Report NE-153).

Ballabeni, P., M. Wlodarczyk, and M. Rahier. 2001. Does enemy free space for eggs contribute to a leaf beetle's oviposition preference for a nutritionally inferior host plant? Funct. Ecol. 15: 318-324.

Barron, A. B. 2001. The life and death of Hopkins' host selection principle. J. Insect Behav. 14(6): 725-737.

Beldade, P. and P. M. Brakefield. 2002. The genetics of evo-devo of butterfly wing patterns. Nature Rev.(Genetics) 3: 442-452.

Berenbaum, M. R. 1981. An oviposition "mistake" by *Papilio glaucus* (Papilionidae) J. Lepid. Soc. 35: 75.

Bergman, A. & M. L. Siegal. 2003. Evolutionary capacitance as a general feature of complex gene networks. Nature 424: 549-552.

Bernays, E. A. 1991. Plasticity and the problem of choice in food selection. Ann. Entom. Soc. Am. 92: 944-951.

Bernays, E. A. 2001. Neural limitations in phytophagous insects: Implications for diet breadth and evolution of host affiliation. Annu. Rev. Entomol. 46: 703-727

Bernays, E.A. and M.R. Weiss. 1996. Induced food preference in caterpillars: the need to identify mechanisms. Entomol. Exp. Appl. 78: 1-8.

Bjorksten T.A. and Hoffmann AA. 1998. Plant cues influence searching behaviour and parasitism in the egg parasitoid *Trichogramma* nr. Brassicae. Ecol. Entomol. 23(4): 355-362

Bossart, J. L. 2003. Covariance of preference and perforrmance on normal and novel hosts in a locally monophagous and locally polyphagous butterfly population. Oecologia 135: 477-486.

Bossart, J. L. and J. M. Scriber. 1995a. Genetic variation in oviposition preferences in the tiger swallowtail butterfly: interspecific, interpopulation, and interindividual comparisons. pp. 183-194 *In* J.M. Scriber, Y. Tsubaki, and R.C. Lederhouse [eds.] The Swallowtail Butterflies: Their Ecology and Evolutionary Biology. Scientific Publ., Inc., Gainesville, FL.

Bossart, J.L. and J. M. Scriber. 1995b. Maintenance of ecologically significant genetic variation in the tiger swallowtail butterfly through differential selection and gene flow. Evolution 49: 1163-1171.

Bossart, J.L. and J. M. Scriber. 1999. Preference variation in the polyphagous tiger swallowtail butterfly (Lepidoptera; Papilionidae). Env. Entomol. 28: 628-637.

Braby, M. F. 1994. The significance of egg size variation in butterflies in relation to host plant quality. Oikos 71: 119-129.

Brower, L. P. 1958. Larval foodplant specificity in butterflies of the *Papilio glaucus* group. Lepid. News 12: 103-114.

Brower, L. P. 1959. Speciation in butterflies of the *Papilio glaucus* group. II, Ecological relationships and and interspecific sexual behavior. Evolution 13: 212-228.

Brown, K. S. Jr., A. J. Damman, and P. Feeny. 1981. Troidine swallowtails (Lepidopteraa: Papilionidae) in southeastern Brazil: Natural history and foodplant relationships. J. Res. Lep. 19: 199-226.

Carter, M. and P. Feeny. 1999. Host plant chemistry influences choice of the spicebush swallowtail butterfly, *Papilio troilus*. J. Chem. Ecol. 25: 1999-2009.

Carter, M., P. Feeny, and M. Haribal.1999. An oviposition stimulant for the spicebush swallowtail butterfly, *Papilio troilus* (Lepidoptera: Papilionidae) from leaves of *Sassafras albidum* (Lauraceaae). J. Chem. Ecol. 25: 1233-1245.

Chew, F. S. 1977. Coevolution of pierid butterflies and their cruciferous foodplants. II. The distribution of eggs on potential foodplants. Evolution 31: 568-579.

Cooley, S.S. , R.J. Prokopy, P.T. McDonald, & T.Y. Yang. 1986. Learning in oviposition site selection by *Ceratitus capitata* flies. Entomol. Exp. Appl.40: 47-51.

Corbet, S.A. 1985. Insect chemosensory responses: a chemical legacy hypothesis. Ecol. Entomol. 10: 143- 153.

Courtney, S.P. 1982. Coevolution of pierid butterflies and their cruciferous food plants: IV. Crucifer apparency and *Anthocharis cardamines* oviposition. Oecologia 52: 258-265.

Courtney, S.P. and T.T. Kibota. 1990. Mother doesn't know best: selection of hosts by ovipositing insects. Pp161-188. *In* E.A. Bernays [ed.] Insect-Plant Interactions. Vol 2. CRC Press, Boca Raton, FL.

Cunningham, J.P., M.F.A. Jallow, and D.J. Wright. 1998. Learning in host selection in *Helicoverpa armigera* (Hubner) (Lepidoptera: Noctuidae). Anim. Behav. 55: 227-234.

Cunningham, J.P., S.A. West, and M. Zalucki. 2001. Host selection in phytophagous insects: a new explanation for learning in adults. Oikos 95: 537-543.

Dahlgaard, J., E. Hasson, and V. Loeschcke. 2001. Behavioral differentiation in oviposition activity in *Drosophila buzzatii* from highland and lowland populations in Argentina: Plasticity or thermal adaptation? Evolution 55: 738-747.

Danks, H. 1994. Insect life-cycle polymorphism: theory, evolution and ecological consequences for seasonality and diapause control. Kluwer Academic Publ., Dordrecht, Netherlands.

Danks, H. 1999. Life cycles in polar arthropods- flexible or programmed? Eur. J. Entomol. 96: 83-103.

Damman, H. and P. Feeny. 1988. Mechanisms and consequences of selective oviposition by the zebra swallowtail butterfly. Anim. Behav.. 36: 563-573.

Del Campo, M. L, C. I. Miles, F. C. Schroeder, C. Mueller, R. Booker, and J. A. Renwick. (2001). Host recognition by the tobacco hornworm is mediated by a host plant compound. Nature 411: 186 – 189

DeMott, W.R., and A.J. Tessier. 2002. Stoichiometric constraints vs. algal defenses: Testing mechanisms of zooplankton food limitation. Ecology 83(12): 3426-3433

Denno, R., M.S. McClure, and J.R. Ott. 1995. Interspecific interactions in phytophagous insects: competion reexamined and resurrected. Annu. Rev. Entomol. 40: 297-331.

Dewitt, J.D. A.. Sih, and D.S. Wilson. 1998. Costs and limits of phenotypic plasticity. Trends Ecol. Evol. 13: 77-81.

Edmands, S. 1999. Heterosis and outbreeding depression in interpopulation crosses spanning a wide range of divergence. Evolution 53: 1757-1768.

Ellers, J. and M. Jervis. 2003. Body size and the timing of egg production in parasitoid wasps. Oikos 102(1): 164-172

Elser. J. J., W. F. Fagan, R. F. Denno, D. R. Dobberfuhl, A. Folarin, A. Huberty, S. Interlandi, S. S. Kilham, E. McCauley, K. L. Schulz, E. H. Siemann,and R. W. Sterner. 2000. Nutritional constraints in terrestrial and freshwater food webs. Nature 408: 578-580

Feeny, P. 1970. Seasonal changes in oak leaf tannins and nutrients as a cause of spring feeding by winter moth caterpillars. Ecology 51: 565-581.

Feeny, P. 1995. Ecological opportunism and chemical constraints on the host associations of swallowtail butterflies. Pp 9-15. *In* Scriber, J.M., Y. Tsubaki, and R.C. Lederhouse [eds].1995. Swallowtail butterflies: Their ecology and evolutionary biology. Scientific Publ., Gainesville, FL.

Feeny, P., L. Rosenberry, and M. Carter. 1983. Chemical aspects of oviposition behavior in butterflies. pp. 27-76. *In* S. Ahmad [ed.]. 1983. Herbivorous insects: Host seeking behavior and mechanisms. Academic Press, NY

Fitt, G.P. 1986. The influence of shortage of hosts on the specificity of oviposition behavior in species of *Dacus* (Diptera: Tephrididae) Physiol. Entomol. 11: 133-143.

Frankfater, C. R. and J. M. Scriber. 1999. Chemical basis for host recognition by two oligophagous swallowtail butterflies, *Papilio troilus* and *Papilio palamedes*. Chemoecology 9:127-132.

Frankfater, C. R. and J. M. Scriber. 2002. Oviposition deterrency of redbay (Persea borbonia) leaf extracts to a sympatric generalist herbivore, *Papilio glaucus* (Papilionidae). Holarct. Lepid. 7(2): 33-38.

Fry, J.D. 1996. The evolution of host specialization: are trade-offs overrated? Am. Nat. 148: S84- S107

Graves, S.D. and A. M. Shapiro. 2003. Exotics as host plants of the California butterfly fauna. Biol. Conserv. 110: 413-4333.

Groeters, F.R. and H. Dingle. 1988. Genetic and maternal influences on the life history plasticity in milkwed bugs (*Oncopeltus*): response to temperature. J. Evol. Bio. 1: 317-333.

Grossmueller, D. W. and R. C. Lederhouse. 1985. Oviposition site selection; An aid to rapid growth and development in the tiger swallowtail, *Papilio glaucus*. Oecologia 66: 68-73.

Harris, M. O. and S. Rose. 1990. Chemical, color, and tactile cues influencing ovipositional behavior of the Hessian fly (Diptera: Cecidomyiidae) Environ. Entomol. 19: 303-308.

Heimpel, G. E. and J.A. Rosenheim. 1998. Egg limitation in parasitoids - A review of the evidence and a case study. Biol. Control 11: 160-168.

Hilker, M. and T. Meiners. 2002. Induction of plant responses to oviposition and feeding by herbivorous arthropods: a comparison. Entomol. Exp. Appl.104: 181-192.

Horton, D.R. and J. L. Krysan. 1991. Host acceptance behavior of pear psylla (Homoptera, Psyllidae) affected by plant species, host deprivation, habituation, and egg load. Ann. Entomol. Soc. Am. 84(6): 612-627

Huang, X. P. & J. A. A. Renwick. 1995. Cross habituation to feeding deterrents and acceptance of a marginal host plant by *Pieris rapae* larvae. Entomol. Exp. Appl.76: 295-302.

Jaenike, J. 1982. Environmental modification of oviposition behavior in *Drosophila*. Am. Nat. 119: 784- 802.

Jaenike, J. 1983. Induction of host preference in *Drosophila melanogaster*. Oecologia 58: 320-325.

Jaenike, J. 1990. Host specialization in phytophagous insects. Ann. Rev. Ecol. Syst. 21: 243-273.

Jaenike, J. & D.R. Papaj. 1992. Behavioral plasticity and patterns of host use by insects. Pp 245-264. *In* B.D. Roitberg and M.B. Isman [eds.] Insect Chemical Ecology: an Evolutionary Approach. Chapman and Hall NY.

Jallow, M.F.A. and M.P. Zalucki. 1998. Effects of egg load on the host-selection behaviour of *Helicoverpa armigera* (Hubner) (Lepidoptera : Noctuidae). Aust. J. Zool. 46(3): 291-299

Janz, N. 1998. Sex-linked inheritance of host-plant specialization in a polyphagous butterfly. Proc R Soc Lond B Biol Sci. 265: 1675-1678.

Janz, N. 2003. Sex linkage of host plant use. Pp229-239. *In* C.L. Boggs, W.B. Watt, and P.R. Ehrlich [eds.] Butterflies: Ecology and evolution taking flight. Univ. Chicago Press, Chicago, IL.

Jermy, T. 1984. Evolution of insect plant relationships. Am. Nat. 124: 609-630.

Jermy, T. 1988. Can predation lead to narrow food specialization in phytophagous insects? Ecology 69: 902-904.

Jones, R.E. 1977. Movement patterns and egg distribution in cabbage butterflies. J. Anim. Ecol. 46: 195- 212.

de Jong, P.W. and J.K. Nielsen. 2002. Host plant use of *Phyllotreta nemorum*: do coadapted gene complexes play a role? Entomol. Exp. Appl.104: 207-215.

Kareiva, P. 1985. Finding and losing host plants by *Phyllotreta*: patch size and surrounding habitat. Ecology 66: 1809-1816.

Kelly, D., and A. Monks. 2003. Motivation models fail to explain oviposition behaviour in the diamondback moth. Physiol. Entomol. 28 (3): 199-208.

Kingsolver, J.G. and R.B. Huey. 1998. Evolutionary analyses of morphological and physiological plasticity in thermally variable environments. Amer. Zool. 38: 545-560.

Landolt, P.J. and O. Molina. 1996. Host-finding by cabbage looper moths (Lepidoptera: Noctuidae): learning of host odor upon contact with host foliage. J. Insect Behavior 9: 899-908.

Larsson, S. & B. Ekbom. 1995. Oviposition mistakes in herbivorous insects: confusion or a step towards a new host plant? Oikos 72: 155-160.

Lederhouse, R.C. and J.M. Scriber. 1987. The ecological significance of post-mating decline in egg viability in the tiger swallowtail, *Papilio glaucus* L. (Papilionidae) J. Lep. Soc. 41: 83-93.

Lederhouse, R.C., M. Finke and J.M. Scriber. 1982. The contributions of larval growth and pupal duration to protandry in the black swallowtail butterfly, *Papilio polyxenes*. Oecologia 53: 296- 300.

Li, X., M. R.Schuler and M. R. Berenbaum. 2002a. Jasmonate and salicylate induce expression of cytochrome P450 genes in insects. Nature 419:712-715.

Li, X., R.. A. Petersen, M. R. Schuler and M. R. Berenbaum. 2002b. CYP6B cytochrome P450 monooxygenases from *Papilio canadensis* and *Papilio glaucus:* potential contributions of sequence divergence to host plant associations. Insect Mol. Biol.11: 343-351.

Mark, G.A. 1982. Induced oviposition preference, periodic environments, and demographic cycles in the bruchid beetle, *Callosobruchus maculatus* Fab. Entomol. Exp. Appl.21: 155-160.

Mattson, W.J. and J.M. Scriber. 1987. Nutritional ecology of insect folivores of woody plants: water, nitrogen, fiber, and mineral considerations. pp. 105-146. In F. Slansky, Jr. and J.G. Rodriguez [eds.] 1987. Nutritional Ecology of Insects, Mites, and Spiders. John Wiley, NY.

Mayhew, J.P. 1997. Adaptive patterns of host-plant selection by phytophagous insects. Oikos 79: 417- 428.

Mayhew, J. P. 2001. Herbivore host choice and optimal bad motherhood. Trends Ecol. Evol. 16: 165-167.

Miller, J. R. and K.L. Strickler. 1984. Finding and accepting host plants. Pp 127-158. In J.W. Bell and R. T. Carde [eds.] Chemical Ecology of Insects. Chapman and Hall, NY.

Minkenberg, O.P., M. Tatar, & J.A. Rosenheim. 1992. Egg load as a major source of variability in insect foraging and oviposition behavior. Oikos 65: 134-143.

Monks, A. and Kelly, D. 2003. Motivation models fail to explain oviposition behaviour in the diamondback moth. Physiol. Entomol. 28(3):199-208

Mousseau, T.A., B. Sinervo, J. Endler. 2000. Adaptive genetic variation in the wild. Oxford University Press, NY 265pp.

Mousseau, T.A. and D.A. Roff. 1987. Natural selection and the heritability of fitness components. Heredity 59: 181-197.

Nation, J. L. 2002. Insect Physiology and Biochemistry. CRC Press, Boca Raton.

Niemala, P. and W. J. Mattson. 1996. Invasion of North American forests by European phytophagous insects: Legacy of the European crucible? BioScience 46: 741-753.

Nitao, J.N. 1995. Evolutionary stability of swallowtail adaptations to plant toxins. Pp 39- 52 *In* J.M. Scriber, Y. Tsubaki, and R.C. Lederhouse [eds] 1995. Swallowtail Butterflies; Their Ecology and Evolutionary Biology. Scientific Publ, Gainesville,FL,

Nylin, S. 1988. Host plant specialization and saeasonality in a polyphagous butterfly *Polygonia c-album* (Lepidoptera: Nymphalidae) Biological Jourrnal of the Linnean Society 47: 301-323.

Nylin, S. 1994. Seasonal plasticity and life cycle adaptations in butterflies. Pp 41-67 In (Danks, H. ed) Insect life-cycle polymorphism: theory, evolution and ecological consequences for seasonality and diapause control. Kluwer Academic Publ., Dordrecht, Netherlands.

Nylin, S. and K. Gotthard. 1998. Plasticity in life history traits. Annu. Rev. Entomol. 43: 633-83.

Nylin, S., N. Janz, and N. Wedell. 1996. Oviposition plant preference and offspring performance in the comma butterfly: correlations and conflicts. Entomol. Exp. Appl.80: 141-144.

Ohsaki, N. & Y. Sato. 1999. The role of parasitoids in the evolution of habitat and larval food plant preference by three species of *Pieris* butterflies. Res. Popul. Ecol.(Kyoto) 41:107-119.

Odendal F. J. and M. D. Rausher. 1990. egg load influences search intensity, host selectivity and clutch size in *Battus philenor* butterflies. J. Insect Behav. 3: 183-193.

Papaj, D.R. 2000. Ovarian dynamics and host use. Annu. Rev. Entomol. 45: 423-448.

Papaj, D.R. and A.C. Lewis [eds]. 1993. Insect Learning: Ecological and Evolutionary Perspectives. Chapman & Hall, Inc. 398pp.

Papaj, D.R. and R.J. Prokopy. 1989. Ecological and evolutionary aspects of learning in phytophagous insects. Annu. Rev. Entomol. 34: 315-350.

Papaj, D.R. and M. D. Rausher. 1987. Components of conspecific host discrimination in the butterfly *Battus philenor*. Ecology 68: 245-253.

Parry, D., R.A. Goyer, and G.J. Lenhard. 2001. Macrogeographic clines in fecundity, reproductive allocation, and offspring size of the forest tent caterpillar *Malacosoma disstria*. Ecol. Entomol. 26: 281-291.

Partridge, L. & J.A. Coyne. 1997. Bergman's rule in ectotherms: is it adaptive? Evolution 51: 632-635.

Piersma, T. & J. Drent. 2003. Phenotypic flexibility and the evolution of organism design. Trends Ecol. Evol.. 18: 228-233.

Pigliucci, M. 2001. Phenotypic plasticity. John Hopkins University Press.

Pigliucci, M. and C.J. Murren. 2003. Perspective: Genetic assimilation and a possible evolutionary paradox: Can macroevolution sometimes be so fast, as to pass us by? Evolution 57: 1455-1464.

Poulton, E.B. 1892. Further experiments upon colour relations between certain Lepidopterous larvae, pupae, cocoon and imagines and their surroundings. Trans. Entomol. Soc. Lond. 1892: 293-487.

Prokopy, R.J., A. Averill, S.S. Cooky, and C.A. Roitberg. 1982. Associative learning in egg-laying site selection by apple maggot flies. Science 218: 76-77.

Prokopy, R.J., B.D. Roitberg, and R.I. Vargas. 1994. Effects of egg load on finding and acceptance of host fruit in *Ceratitus capitata* flies. Physiol. Entomol. 199: 124-132.

Raffa, K.F., N.P. Havill, and E.V. Nordheim. 2002. How many choices can your test animal compare effectively? Evaluating a critical assumption of behavioral preference tests. Oecologia 133: 422 - 429

Rausher, M.D. 1978. Search image for leaf shape in a butterfly. Science 200: 1071-1073.

Rausher, M.D. 1983. Conditioning and genetic variation as causes of individual variation in the oviposition behavior of the tortoise beetle, *Deloyata guttata*. Anim. Behav. 31: 743-747.

Ray, S. 1999. Survival of olfactory memory through metamorphosis in the fly *Musca domestica.* Neurosci. Lett., 259: 37-40.

Redman, A. and J.M. Scriber. 2000. Competition between gypsy moth, *Lymantria dispar*, and the northern tiger swallowtail, *Papilio canadensis:* interactions mediated by host plant chemistry, pathogens, and parasitoids. Oecologia 125 (2): 218-228.

Renwick, J. A. A. 1990. Oviposition stimulants and deterrents. Pp151-160. *In* E.D. Morgan and N.B. Mandava [eds] 1990. Handbook of natural pesticides: Insect attractants and repellants. CRC Press, Boca Raton.

Renwick, J. A. A. 2002. The chemical world of crucivores: lures, treats, and traps. Entomol. Expt. Appl. 104: 35-42.

Renwick, J. A. A. & F. Chew. 1994 Oviposition behavior in Lepidoptera. Annu. Rev. Entomol. 39: 377- 400.

Renwick, J. A. A. and K. Lopez. 1999. Experience-based food consumption by larvae of Pieris rapae: addiction to glucosinolates. Entomol. Exp. Appl. 91: 51-58.

Roff, D. A. 1996. The evolution of threshold traits in animals. Q. Rev. Biol. 71: 3-35.

Roff, D.A. 2002. Life history evolution. Sinauer, Sunderland, MA. 527 pp.

Rosenheim, J.A. 1996. An evolutionary argument for egg limitation. Evolution 50: 2089-2094.

Rosenheim, J. A. 1999. Characterizing the cost of oviposition in insects: a dynamic model. Evol. Ecol. 13 (2): 141-165 1999

Rosenheim, J. A., G. E. Heimpel, and M. Mangel. 2000. Egg maturation, egg resorption and the costliness of transient egg limitation. Proc. R. Soc. Lond. B Biol. Sci. 267:1565-1573.

Schemske, D.W. 2000. Understanding the origin of species. Evolution 54: 1069-1073.

Scheirs, J., L. De Bruyn, and R. Verhagen. 2000. Optimization of adult performance determines host choice in a grass miner. Proc. R. Soc. Lond. B Biol. Sci. 267: 2065-2069.

Scheirs, J. and L De Bruyn. 2002. Integrating optimal foraging and optimal oviposition theory in plant- insect research. Oikos 96: 187-191.

Schlichting, C.D. and M. Pigliucci. 1998. Phenotypic evolution: A reaction norm perspective. Sinauer, Sunderland MA

Schmitz, O.J., F.R. Adler, and A.A. Agrawal. 2003. Linking individual-scale trait plasticity to community dynamics (special feature) Ecology 84: 1081- 1082.

Schoonhoven, L.M., T. Jermy, & J.A.A. van Loon. 1998. Insect-plant biology: From physiology to evolution. Chapman & Hall, NY

Schultz, J. C. 2002. Shared signals and the potential for phylogenetic espionage between plants and animals. Integr. Comp. Biol. 42 (3): 454-462.

Scriber, J. M. 1984a. Larval foodplant utilization by the World Papilionidae (Lepidoptera): Latitudinal gradients reappraised. Tokurana (Acta Rhopalocerologica) 2: 1-50.

Scriber, J. M. 1988. Tale of the tiger: Beringial biogeography, bionomial classification, and breakfast choices in the *Papilio glaucus* complex of butterflies. pp. 240-301 In K.C. Spencer [ed.]Chemical Mediation of Coevolution. Academic Press. NY.

Scriber, J. M. 1993. Absence of behavioral induction in oviposition preference of *Papilio glaucus* (Lepidoptera: Papilionidae). Gt. Lakes Entomol. 26: 81-95.

Scriber, J. M. 1994. Climatic legacies and sex chromosomes: latitudinal patterns of voltinism, diapause size and host-plant selection in 2 species of swallowtail butterflies at their hybrid zone. pp. 133- 171. In H.V. Danks [ed.] 1994. Insect life-cycle polymorphism: theory, evolution and ecological consequences for seasonality and diapause control. Kluwer Academic Publ., Dordrecht, Netherlands

Scriber, J. M. 1995. Diversity of Swallowtails: An overview of taxonomic and distributional latitude. pp. 3-8. In J.M. Scriber, Y. Tsubaki, and R.C. Lederhouse [eds.] The Swallowtail Butterflies: Their Ecology and Evolutionary Biology. Scientific Publishers, Inc., Gainesville, FL.

Scriber, J. M. 1996a. Tiger tales: Natural history of native North American swallowtails. Am. Entomol. 42:19-32.

Scriber, J. M. 1996b. A new cold pocket hypothesis to explain local host preference shifts in *Papilio canadensis*. 9th Intern. Symp. Insects and Host Plants. Entomol. Exp. Appl.80: 315-319.

Scriber, J. M. 1996c. Generalist North American silkmoth, *Antherea polyphemus*, grows more efficiently and rapidly than its specialized relative, *A. pernyi* (Lepidoptera: Saturniidae) Holarct. Lepid. 3:23-30.

Scriber, J. M. 2002a. The evolution of insect-plant relationships; Chemical constraints, coadaptation and concordance of insect/plant traits. Entomol. Exp. Appl. 104: 217-235

Scriber, J. M. 2002b. Latitudinal and local geographic mosaics in host plant preferences as shaped by thermal units and voltinism in *Papilio* spp. (Lepidoptera) . Eur. J. Entomol. 99: 225- 239.

Scriber, J.M. and S. Gage. 1995. Pollution and global climate change: Plant ecotones, butterfly hybrid zones, and biodiversity. pp. 319-344. In J.M. Scriber, Y. Tsubaki, and R.C. Lederhouse [eds.] The Swallowtail Butterflies: Their Ecology and Evolutionary Biology. Scientific Publishers, Inc., Gainesville, FL.

Scriber, J.M. and J. Hainze. 1987. Geographic variation in host utilization and the development of insect outbreaks. pp. 433-468. In (P. Barbosa and J.C. Schultz, eds.) Insect Outbreaks: Ecological and Evolutionary Processes. Academic Press, NY.

Scriber, J.M. and F. Slansky, Jr. 1981. The nutritional ecology of immature insects. Annu. Rev. Entomol. 26: 183-211.

Scriber, J.M., B.L. Giebink and D. Snider. 1991a. Reciprocal latitudinal clines in oviposition behavior of *Papilio glaucus* and *P. canadensis* across the Great Lakes hybrid zone: possible sex-linkage of oviposition preferences. Oecologia 87: 360-368.

Scriber, J.M. and R.C. Lederhouse. 1992. The thermal environment as a resource dictating geographic patterns of feeding specialization of insect herbivores. pp 429-466. In M.R. Hunter, T. Ohgushi and P.W. Price [eds.] Effects of Resource Distribution on Animal-Plant Interactions. Academic Press.

Scriber, J. M., R. C. Lederhouse, and L. Contardo. 1975. Spicebush, *Lindera benzoin*, a little known foodplant of *Papilio glaucus* (Papilionidae). J. Lepid. Soc. 29(1): 10-14

Scriber, J.M., R.C. Lederhouse and R. Hagen. 1991b. Foodplants and evolution within the *Papilio glaucus* and *Papilio troilus* species groups (Lepidoptera: Papilionidae). pp. 341-373. In P.W. Price, TM. Lewinsohn, G.W. Fernandes and W. W. Benson [eds.] Plant-Animal Interactions: Evolutionary Ecology in Tropical and Temperate Regions. John Wiley, NY.

Scriber, J.M., K. Weir, D. Parry, and J. Deering. 1999. Using hybrid and backcross larvae of *Papilio canadensis* and *P. glaucus* to detect induced chemical resistance in hybrid poplars experimentally defoliated by gypsy moths. Entomol. Exp. Appl. 91:233-236.

Shapiro, A.M. 1976. Seasonal polyphenism. Evol. Biol. 9: 259-333.

Singer, M.C. 1983. Determinants of multiple host use by a phytophagous insect population. Evolution 37: 389-403.

Singer, M.C. 2003. Spatial and temporal patterns of checkerspot butterfly-host plant association: the diverse roles of oviposition preference, pp 207-228. In C.L. Boggs, W.B. Watt, and P.R. Ehrlich [eds] 2003. Butterflies: Ecology and evolution taking flight. Univ. Chicago Press, Chicago, IL

Singer, M.C, D. Ng, D. Vasco, and C.D. Thomas.1988. Heritability of oviposition preference and its relationship to offspring performance within a single insect population. Evolution 42: 977-985.

Singer, M.C, D. Ng, D. Vasco, & C.D. Thomas. 1992. Rapidly evolving associations among oviposition preferences fail to constrain evolution of insect diet. Am. Nat. 193: 9-20.

Slansky, F. and J.M. Scriber. 1985. Food consumption and utilization, pp. 87-163. In G.A. Kerkut and L.I. Gilbert [eds.] 1985. Comprehensive Insect Physiology, Biochemistry, and Pharmacology Volume 4. Permagon Press, Oxford, ENGLAND.

Spencer, J. L., M. P. Candolfi, G.L., J.E. Keller, and J.R. Miller. 1995. Onion fly, *Delia antiqua*, oviposition and mating as influenced by insect age and dosage of male reproductive tract extract. J. Insect Behav. 8: 617-635.

Stanton, M. L. 1984. Short-term learning and searching accuracy of egg-laying butterflies. Anim. Behav. 32: 33-40.

Stearns, S. 1989. The evolutionary significance of phenotypic plasticity. BioScience 39: 436-445.

Stearns, S. 2003. Safeguards and spurs. Nature 424: 501-504.

Stearns, S., G. deJong, and R. Newman. 1991. The effects of phenotypic plasticity on genetic correlations. Trends Ecol. Evol. 6: 122-126.

Stephens, D.W. 1989. Variance and the value of information. Am. Nat. 134(1): 128- 140

Sterner, R.W. and J. J. Elser 2002. Ecological Stoichiometry; The biology of elements from molecules to the biosphere. Princeton University Press, NJ

Straatmaan, R. 1962. Notes on certain Lepidoptera ovipositing on plants which are toxic to their larvae. J. Lepid. Soc. 16: 99-103.

Szentesi, A. & T. Jermy.. 1990. The role of experience in host choice by phytophagous insects, pp 39-74 In E.A. Bernays [ed.) Insect-Plant Interactions Vol. 2. CRC Press, Boca Raton.FL.

Thompson, J.N. 1988a. Evolutionary ecology of the relationship between oviposition preference and performance of offspring in phytophagous insects. Entomol. Exp. Appl.47: 3-14.

Thompson, J.N. 1988b. Variation in preference and specificity in monophagous and oligophagous swallowtail butterflies. Evolution 42: 118-128.

Thompson, J.N. 1988c. Evolutionary genetics of oviposition preference in swallowtail butterflies. Evolution. 42: 1223-1234.

Thompson, J.N. 1993. Preference hierarchies and the origin of geographic specialization in host use in swallowtail butterflies. Evolution 47: 1585-1594.

Thompson, J.N. 1994. The coevolutionary process. Univ. Chicago Press, IL

Thompson, J.N. 1998. The evolution of diet breadth: monophagy and polyphagy in swallowtail butterflies. J. Evol. Biol. 11: 563-578.

Thompson, J.N. and O. Pellmyr. 1991. Evolutionary genetics of oviposition behavior and host preferences in Lepidoptera. Annu. Rev. Entomol. 36: 65-69.

Tully, T., V. Cambiazo, and L. Kruse. 1994. Memory through metamorphosis in normal and mutant Drosophila. J. Neurosci. 14: 68-74.

Van Emden, H.F., B. Sponagl, E. Wagner, T. Baker, S. Ganguly, and S. Douloumpka. 1996. Hopkins host selection principal, "another nail in its coffin. Physiol. Entomol. 21: 325-328.

Via, S. 1990. Ecological genetics and host adaptation in herbivorous insects: the experimental study of evolution in natural and agricultural systems. Annu. Rev. Entomol. 35: 421-446.

Via, S. 2001. Sympatric speciation in animals: The ugly duckling grows up. Trends Ecol. & Evol. 16: 381-390.

Via, S., R. Gomulkiewicz, G. DeJong, and S.M. Schiener, C.D. Schlichting, and P.H. van Tienderen. 1995. Adaptive phenotypic plasticity: concerns and controversy. Trends Ecol. Evol. 10: 212- 217

Via, S. and D.J. Hawthorne. 2002. The genetic architecture of ecological specialization: correlated gene effects on host use and habitat choice in pea aphids. Am. Nat. 159: S76-S88.

Vinson, S.B., C.S. Barfield, and R.D. Henson.1977. Oviposition behavior of *Bracon mellitor*, a parasitoid of the boll weevil (*Anthonomus grandis*) II. Associative learning. Physiol. Entomol. 2: 157-164.

Wardle, A.R. and J. H. Borden. 1985. Age-dependent associative learning by *Exeristes roborator* (F.) (Hymenoptera: Ichneumonidae) Canad. Entomol. 117: 605-616.

Wehling, W. F., and J. N. Thompson. 1997. Evolutionary conservatism of oviposition preference in a widespread polyphagous insect herbivore, *Papilio zelicaon*. Oecologia 111:209-215

Weiss, M. R. and D. R. Papaj. 2003. Colour learning in two behavioral contexts: how much can a butterfly keep in mind? Anim. Behav. 65: 425-434.

West-Eberhardt, M. J. 1989. Phenotypic plasticity and the origins of diversity. Annu. Rev. Ecol. Syst. 20: 249-278.

West, S. A. abd J. P. Cunningham. 2002. A general model for host plant selection in phytophagous insects. J. Theor. Biol. 214: 499-513

Wiklund, C. 1975. The evolutionary relationship between adult oviposition preferences and larval host plant range in *Papilio machaon* L. Oecologia 18: 185-197.

Wiklund, C. 1981. Generalist vs. specialist oviposition behavior in *Papilio machaon* (Lepidoptera) and functional aspects on the hierarchy of oviposition preferences. Oikos 36: 163-170.

Wiklund, C., S. Nylin, and J. Forsberg.1991. Sex-related variation in growth rate as a result of selection for large size and protandry in a bivoltine butterfly (*Pieris napi* L.) Oikos 60: 241-250.

Wilson, R.S. & C.E. Franklin. 2002. Testing the beneficial acclimation hypothesis. Trends Ecol. Evol. 17: 66-70.

Withers, T.. & L. Barton Brown. 1998. Possible causes of apparently indiscriminate oviposition in host specificity tests using phytophagous insects, pp. 565-571. In M.P. Zaluki, R.A.I. Drew, and G.G. White [eds.] 1998. Vol. 2: Pest Management – Future Challenges, Univ. Queensland, Brisbane, Australia.

Wolf, J.B. & E.D. Brodie. 1998. The coadaptation of parental and offspring characters. Evolution 52: 299-308.

Zonneveld, C. 1996. Being big or emerging early. Protandry and the trade-off between size and emergence in male butterflies. Am. Nat. 147: 946-965.

3

Plasticity in Insect Responses to Variable Chemistry of Host Plants

Meena Haribal[1,2] and J. Alan A. Renwick[1]

[1]*Boyce Thompson Institute, Tower Road, Ithaca, NY 14853, USA*
e-mail: mmh3@cornell.edu
[2]*Current address: Cornell Lab of Ornithology, 159, Sapsucker Woods Road, Ithaca NY 14850, USA.*

Introduction

The importance of plant chemistry in shaping complex relationships between plants and insects has long been, and continues to be recognized (Bernays and Chapman 1994, Schoonhoven et al. 1998). Specific chemicals or groups of chemicals in plants are known to serve as general defense agents against herbivores and pathogens. However, many insects have adapted to these defenses, to the extent that they may use the "defensive" chemicals for host finding and recognition. The ability of an insect to utilize a plant for food, shelter or escape from enemies clearly depends to a large degree on its behavioral and physiological responses to the plant chemistry. Specialist insects may even sequester the potentially toxic compounds to protect themselves from predators, or they may utilize them for other purposes, such as pheromone biosynthesis (Renwick 2001).

One might expect that the dependence of phytophagous insects on plant chemistry could make them particularly vulnerable to changes in this chemistry, which could result from a plant's response to pathogen attack, herbivory, environmental factors, seasonal changes, or genetic variation. Yet many specialist insects have been shown to have at least some capacity to deal with such changes. In fact, one popular theory of coevolution suggests that the ability of insects to deal with the barrage of chemicals produced by plants is part of an on-going arms race responsible for the close associations that we see between certain insects and their host plants (Ehrlich and Raven

1964). Behavioral responses of the insects to plant chemistry may differ as a result of changes in receptor sensitivity, which may be related to experience. Receptor sensitivity is often reduced or otherwise modified as a result of habituation (del Campo et al. 2001) or sensitization (Bernays 1995, Huang and Renwick 1997), and motivation to accept or reject a plant is affected by deprivation, satiation, and general environmental conditions.

The plasticity of insects that allows them to adapt to changing plant chemistry is now becoming more evident. Several good examples have recently emerged, and studies of the chemical relationships between selected lepidopterous insects and their host plants serve to illustrate some of the mechanisms involved. Here we examine specific examples of monophagous and oligophagous lepidopterans that help to illustrate the role of phenotypic plasticity in the dynamics of insect-plant relationships.

Orientation to and Acceptance of Plants by Monophagous Zebra Swallowtail Butterflies

Responses to Contact Cues

Zebra Swallowtails (*Eurytides marcellus*) (Papilionidae) are monophagous insects, producing 2-4 generations annually, while feeding only on plants within the genus *Asimina* (Annonaceae), commonly known as pawpaw. Their oviposition behavior is closely linked to their hosts' phenology. About 6 species of *Asimina* are found in the United States, and except for *Asimina triloba*, which is a medium-sized tree, all of these are small shrubs. The *Asimina* spp. put out a burst of new leaves in the spring and then new growth is only observed if there is damage to the plants by herbivory. The leaves of *Asimina* spp. become either rougher (as in *A. speciosa*) or drier and less fleshy (as in *A. triloba*) as the season progresses. Fresh young leaves are essential for the survival of earlier instars of this butterfly (Damman and Feeny 1988). Chemical studies and bioassays show a complex interaction between plant phenology, plant chemistry (volatile and contact stimulants) and herbivorous insects. 3-caffeoyl*muco*quinic acid (3-C*m*QA) (Fig. 1) from *Asimina triloba* serves as an oviposition stimulant for the zebra swallowtail butterfly (Haribal and Feeny 1998, Haribal et al. 1998). However, the plant also contains flavonols, of which rutin (Fig. 2,1) and nicotiflorine (Fig 2,2) are the major and minor components, respectively. The young leaves have high concentrations of 3-C*m*QA and very low concentrations of flavonoids in early spring. But as the season progresses, the concentration of stimulant decreases and the flavonoid levels increase (Fig. 3) (Haribal and Feeny

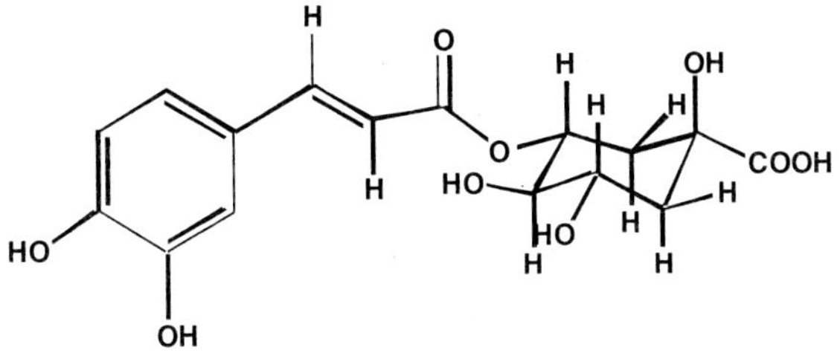
H
O
H
OH
H
H
O
COOH
H
HO
H
HO
HO
H
H
H
OH

Fig. 1 3-Caffeoyl-*muco*-quinic acid: an oviposition stimulant from *A. triloba* for zebra swallowtail butterfly.

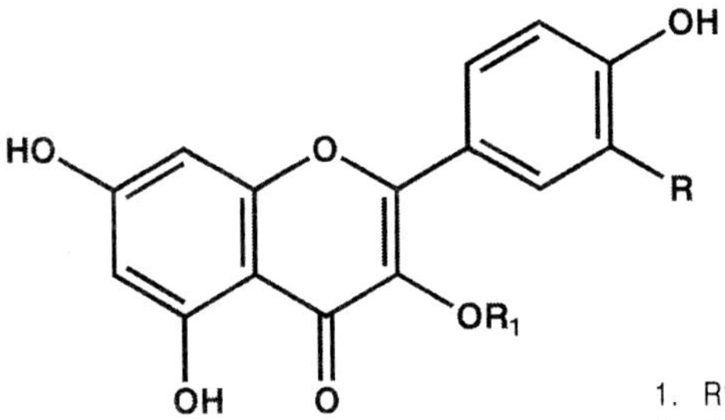
OH
HO
O
R
OR1
OH
O
1. R1 = rutinosyl, R = OH (rutin)
2. R1 = rutinosyl, R = H (nicotiflorine)

Fig. 2 Flavanol glycosides - oviposition deterrents for zebra swallowtail butterflies from *A. triloba*

2003). The butterflies cue in on the relative concentrations of 3-C*m*QA and flavonoids to either accept or reject the host plants; and to avoid ovipositing on those plants that have lower concentrations of stimulants and higher concentrations of flavonoids. Even in the presence of higher concentrations of stimulant, if the concentration of flavonoids is also high they are still deterred from laying eggs (Haribal and Feeny 2003). Thus the zebra swallowtail butterflies have evolved to use this variation in chemistry to assess appropriateness of the plant for their progeny.

Larvae reared on older leaves suffer higher mortality (Fig. 4) when leaves generally have higher concentrations of the flavonoids. We have not yet confirmed whether flavonoids themselves have feeding deterrent activity or if changes in flavonoid concentrations may be indirectly related to other changes in plant chemistry that affect fitness of the larvae. The adult butterflies of the family Papilionidae and some Nymphalidae (such as monarchs) have evolved to respond to flavonoids for oviposition. This response may be either positive, negative or neutral. But response to flavonoids may be evolutionarily ancient. Thus, to avoid feeding on unsuitable plants, the butterflies have also evolved a mechanism whereby most individuals go into diapause as the

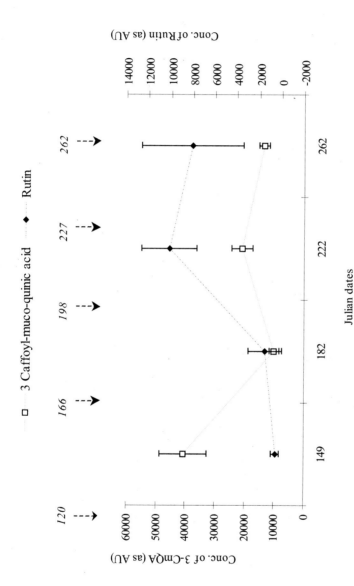

Fig. 3 Seasonal variation in concentrations of 3- caffeoyl-*muco*-quinic acid and flavonols in terminal leaves of *Asimina triloba*, calculated as total area under each peak (AU) in the chromatogram produced by monitoring at 254 nm. The samples were collected on Julian dates 149 (May 29), 182 (July 1), 222 (August 10) and 262 (September 19) of 2000. The vertical dotted arrows show the average eclosion dates of the five generations of butterflies that were reared in our laboratory in 2000.

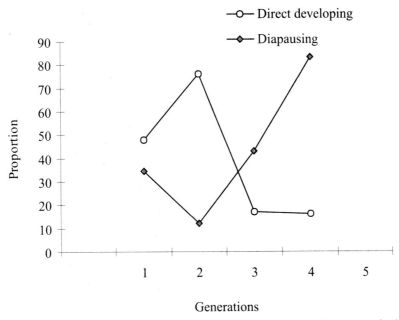

Fig. 4 Proportions (as percentages) of directly developing or diapausing pupae of zebra swallowtail reared in 2000 on *Asimina spp.* for 5 generations from March to September. The colony was started from adult butterflies in spring from diapausing pupae of the previous year's collection in Ocala National Forest (in the last week of March 2000).

season progresses. However, some zebra swallowtail females can tolerate higher amounts of the flavonoids and the progeny will survive on plants of low quality. In Figure 3, we depict the variation in concentration of 3-C*m*QA and flavonoids as the season progresses in the northern latitudes. The dotted arrows represent the mean eclosion dates (Julian) for the butterflies reared in our laboratory. However, three distinct morphological forms of the butterflies (*walshii, telamonides* and *marcellus*) exist. Twenty years of study by Edwards (1868-1897) in the late nineteenth century have shown that in Virginia, the form *walshii* emerges earliest and flies from February to April, the form *telamonides* flies between April and early June and the form *marcellus* generally flies between June and September (Edwards 1897). He also observed that the form *marcellus* does not overlap with the spring brood. Thus butterflies of the later broods may have evolved to cope with this change in the host chemistry, although their survival rate is low (Fig. 4).

In general, the different seasonal forms and the polyphenism in larval, pupal and adult colors is the result of phenotypic plasticity, which is due to the interaction between the environment during development and also, to some extent the genetic background of an individual. Environmental cues for different forms are not fully understood, though hostplant quality, day length and temperature are known to play an important role (Hazel 1995, Greene 1996, Cushman et al. 1994). The ability of insects to respond to different cues provides plasticity to deal with environmentally induced changes in plant chemistry. The studies by Hazel (2002) confirm that the polyphenism in larvae and pupae of Black Swallowtail butterflies is due to environmental factors (i.e phenoplasticity) whereas genetics do not play such an important role (Hazel 2002).

Responses to Volatile Cues

Volatiles are used by many organisms to recognize their hosts (Baur et al. 1993, McCall et al. 1993, Mewis et al. 2002), mates (Borges et al. 1987, Ruther et al. 2000), prey or predators (Dicke et al. 1998, Shimoda and Dicke 2000, Erbilgin and Raffa 2001, Pettersson and Boland 2003). In phytophagous insects, odors are used to locate the host plant for oviposition. Often attractions to host odors are followed by further use of contact cues at the plant surface for final recognition of a site for egg deposition. But not all individuals of a species respond in the same way to available volatiles. In spring in Florida, adult zebra swallowtail butterflies sometimes emerge prior to leaf emergence in their host, *A. speciosa*, and lay eggs on the tiny leaf buds. The lack of visible leaves suggests that the butterflies use volatile cues to locate the plants. Several experiments were conducted to test this assumption. When about 60 naïve butterflies were tested, only 17 females responded to host plant volatiles extracted into hexane (HV), and 6 of these did not lay eggs, although they landed significantly more frequently in the presence of volatiles than on controls (Fig. 5). But more than 50 % of the experienced (either to host plants or to extracts applied to model plants in the experiments) butterflies reacted to host volatiles of *Asimina triloba*, present in a hexane extract (HV) of the foliage (Haribal and Feeny 1998).

When hexane volatiles (HV) were fractionated by MPLC (medium pressure liquid chromatography) into 5 fractions (F1-F5) using hexane, dichloromethane and diethylether combinations of increasing polarity, and were tested in choice assay experiments, fractions F1 to F3 were attractive to the butterflies that had previous experience with the whole hexane volatiles (Fig. 6). F2 was the most active fraction (37% of females tested) for approaches, but was not significantly different from the others for egg-laying responses. A few of the females showed significant preferences for

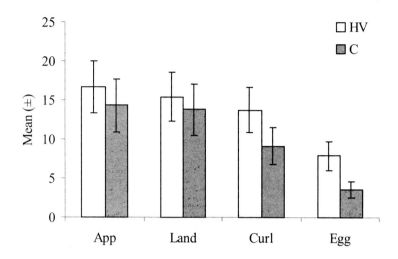

Fig. 5 Mean responses (± SE) of *E. marcellus* to volatiles (3 gle) in the hexane extracts dissolved in hexane (HV) of the foliage of *A. triloba* and to the control (C, solvent hexane). App = Approaches; Land = Landing; Curl = Abdomen Curling and Egg = Egg-laying. The results were analyzed by Wilcoxon's paired rank test (P = <0.001 for approaches and eggs).

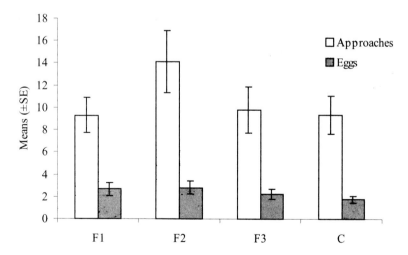

Fig. 6 Mean responses (± SE) (for approaches and eggs) of *E. marcellus* to the F1-F3 MPLC fractions of hexane extract (HV) of *A. triloba* and solvent control (hexane). The results were analyzed by *Friedman's Two-way Rank Analysis* (P = 0.009 and 0.364 for approaches and eggs respectively).

either F1 (15%) or F3 (9%) over F2. But the remaining butterflies had equal preferences for two fractions, either as a combination of F1 and F2 or F2 and F3, and a few even chose the hexane control.

Although all females are initially attracted to the host volatiles in the hexane extract, it appears that they use this experience to memorize some combination of volatiles for recognition of the plants. However, the volatiles used in this way depend on the individual females. This difference in recognition patterns may be a result of different detection thresholds for the different volatile components of the extract.

From these results we can conclude that there is considerable variation among individual females in their responses to volatile cues. A small percentage of naive females respond to volatile cues alone, and only a few of these lay eggs in the absence of contact stimulants. Experienced females may recognize components of the volatile profile of a host plant leading to enhanced egg laying (Haribal and Feeny 1998). However, there is no one magic compound or group of compounds in volatiles that attracts the butterflies to plants, and individual butterflies may have variable thresholds for detection of different components in the volatile blends. Although the role of volatiles in host choice may indeed be significant, the final decision to oviposit is most often dependent on responses to a balance of contact cues.

A number of studies in the literature show that many insect species use volatiles as cues for various functions. But for the first time we show that individuals of a species may differ in their choice of volatile compounds that they use to locate and recognize their host plant. Furthermore, this variation among individuals may depend to a large extent on their previous experience. This has immense ecological implications, as plant chemistry, and particularly release of volatiles, varies even within a single day (Gouinguené and Turlings 2002, Hilker and Meiners 2002, Pott et al. 2002). We do not yet know whether these responses to volatiles are based on environmental cues or genetic traits. However, Changes in volatiles are strongly induced by environmental factors. So, phenoplasticity is likely to be one of the factors that enable insects to respond differently to an array of chemical cues.

Host Recognition by Oligophagous Monarch Butterflies

Contact cues from *Asclepias* spp.

In contrast to the zebra swallowtails, monarch butterflies (*Danaus plexippus*, Fam: Nymphalidae) are oligophagous insects, and their transitory life cycle exposes them to highly variable conditions. In the late fall, adults migrate from the northern United States and Canada to Mexico or Southern California, where they remain reproductively dormant through the winter. As spring arrives in the southern part of the United States, they become reproductively active and lay eggs at the first opportunity as they start their long migration to the north. Although these overwintered adults probably perish at this stage, the next generations from their eggs move farther north to find new hosts. At this time, they encounter different species of *Asclepias* (Fam: Asclepiadaceae) (Brower1995). We have shown that the chemistry of the different hosts varies considerably, and that the insects use different chemical cues to recognize the different hosts (Haribal and Renwick 1998a, b). In the case of *A. curassavica*, quercetin 3-O-(2″, 6″-O-α-L-dirhamnopyranosyl)- β-D-glucopyranoside (Fig. 7, 1) and quercetin 3-O-(6″-O-β-D-glucopyranosyl)-β-D-galactoside (Fig. 7, 2) served as oviposition stimulants, whereas recognition of *A. incarnata* was based on the presence of

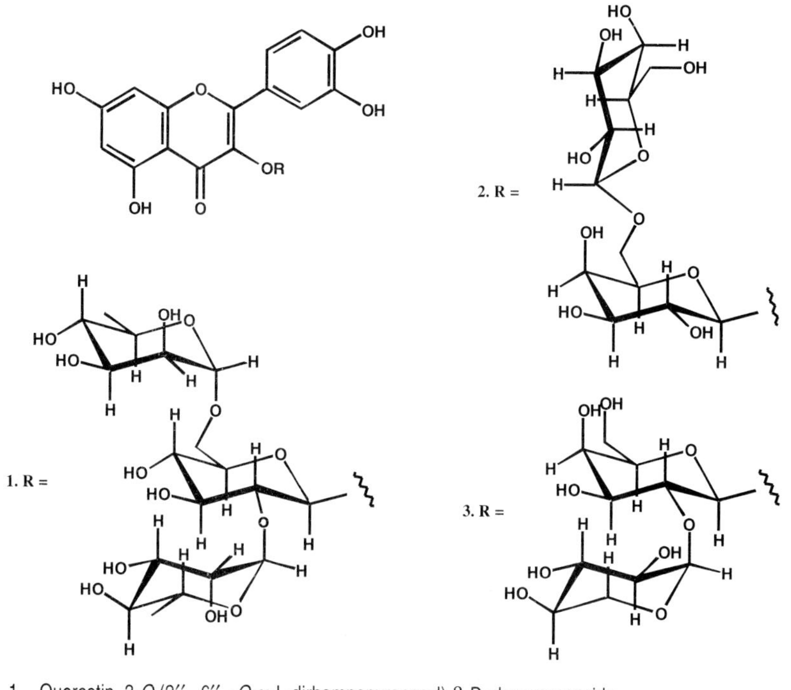

1. Quercetin 3-O-(2″, 6″ -O-α-L-dirhamnopyranosyl)-β-D-glucopyranoside
2. Quercetin 3-O-(6″ -O-β-D-glucopyranosyl)-β-D-galactoside
3.   3-O-(2″, -O-β-D-xlyopyranosyl)-β-D-glucopyranoside

Fig. 7   Oviposition stimulants for monarch butterflies in *A. curassavica.*

quercetin 3-O-(2''-O-β-D-xlyopyranosyl)-?β-D-galactoside (Fig. 7, 3) and quercetin 3-O-β-D-galactoside in the foliage.

Involvement of Multiple Chemoreceptors

Since monarchs must be able to respond to different types of chemicals for host recognition, one might ask how their sensory system is equipped to handle the variable information provided. Sensilla that have been shown to contain receptor neurons sensitive to compounds mediating oviposition are present on antennae, forelegs and midlegs of the butterflies, and the behavior and use of these appendages on different hosts were found to differ significantly (Haribal and Renwick 1998b). On *A. incarnata,* they used forelegs more often than the antennae whereas on *A. curassavica* they used the antennae significantly more often. Furthermore, electrophysiological experiments showed that the contact stimulant receptors on the different appendages have apparently evolved to respond to different chemicals (Baur et al. 1998). Out of 42 receptors (Fig. 8) on antennae, only 11 responded to quercetin 3-O-(6''-O-β-D-glucopyranosyl)-β-D-galactoside (Fig. 7, 2). On the foretarsus they have usually a cluster of nine sensilla associated with each spine (Figs. 9 and 10). Several of these sensilla were tested for sensitivity with isolated stimulants. Overall, responses for the pure compounds were higher than those for solvents, but not as strong as to the fraction containing a mixture of flavonoids. The sensilla on tarsomers 2, 3 and 4 (Fig. 11) of the midleg responded to Quercetin 3-O-(2'',6''-O-α-L-dirhamnopyranosyl)-β-D-glucopyranoside (Fig. 7, 1). During observations of host recognition behavior, the gravid butterflies were often seen to use their midlegs followed by antennae on *A. curassavica.*

Baur et al. (1998) also found that individual females differ in their response to these compounds. Furthermore, all sensilla on a given appendage did not respond to these compounds equally. The overall responses to the flavonoids that were isolated as oviposition stimulants were not as high as those to mixed extracts of flavonoids or to the n-butanol extract of *A. curassavica.* Possible responses of these appendages to the stimulants we isolated from *A. incarnata* have yet to be tested. Recent work in Japan has shown that the ovipositors of many species of butterflies have photoreceptors that are used in oviposition (Arikawa 2001, Arikawa and Takagi 2001). Since these receptors are sensitive to UV, and flavonoids absorb strongly in the UV, ovipositor plates of monarchs were examined. We found no chemoreceptor sensilla, but no attempt was made to identify photoreceptors (Haribal and Renwick 1998b) that theoretically might react

Fig. 8   The tip of the antenna of a Monarch butterfly. (arrow indicates one of the sensilla)

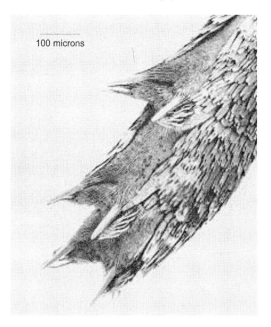

Fig. 9   Spines and sensilla of the foreleg tarsus of the Monarch butterfly.

Fig. 10 Foretarsus spine and sensilla (enlarged 10x) (arrow indicates cluster of sensilla).

Fig. 11 Midleg spines with sensilla. (arrow indicates sensilla between the spines)

to the UV absorption of flavonoids. It appears, therefore, that monarch butterflies have evolved a different strategy to cope with changes in host chemistry and have evolved a method to recognize the different host plants by changing their receptor sensitivities. This may well be a mechanism that is used by other herbivores in dealing with changing environments.

Therefore, it is likely that variable responses may be due to phenoplastic characters of individuals, whereby the use of different receptors for different chemicals allows for responses to environmentally induced changes in plant chemistry.

Host Discrimination by Larvae and Adults of *Pieris* Butterflies

Pierid butterflies, particularly those in the genus *Pieris* (Fam: Pieridae), have been widely studied and used as models for exploring the chemistry of plant-insect relationships. *Pieris* butterflies oviposit exclusively on plants that contain glucosinolates, which have been shown to act as contact cues for oviposition by several species. Cabbage (*Brassica oleracea*) is acceptable for all species of *Pieris*, and has served as a control plant for numerous studies of factors affecting host selection behavior. The common agricultural pests, *Pieris brassicae* and *Pieris rapae* both rely on the generally high concentrations of indole glucosinolates for recognition of *B. oleracea* (Fam: Cruciferae) as hosts (van Loon et al., 1992; Renwick et al., 1992). However, the ubiquitous subspecies of *Pieris napi* are tuned to aliphatic glucosinolates, which are generally prevalent in wild mustards. For example, acceptance of *Erysimum cheiranthoides* for oviposition by *P. napi oleracea* is dependent on the presence of two predominant glucosinolates, glucoiberin and glucocheirolin (Fig. 12). When the behavior of *P. napi* and *P. rapae* was compared, glucoiberin was active only for *P. napi*, but glucocheirolin was somewhat stimulatory for both species. Avoidance of this plant by *P. rapae* could be explained by the presence of cardenolides (erysimoside and erychroside), which act as oviposition deterrents (Sachdev-Gupta et al., 1990). Other examples of crucifers that are avoided by *P. rapae*, but acceptable to *P. napi* have been examined, and the importance of the balance between stimulants and deterrents has been emphasized as a major factor controlling acceptance or rejection of a plant (Renwick and Huang, 1994).

Environmental conditions, previous herbivory and seasonal variation in chemistry can greatly affect the balance between stimulants and deterrents. Growing *Erysimum cheiranthoides* plants under different nutrition regimes

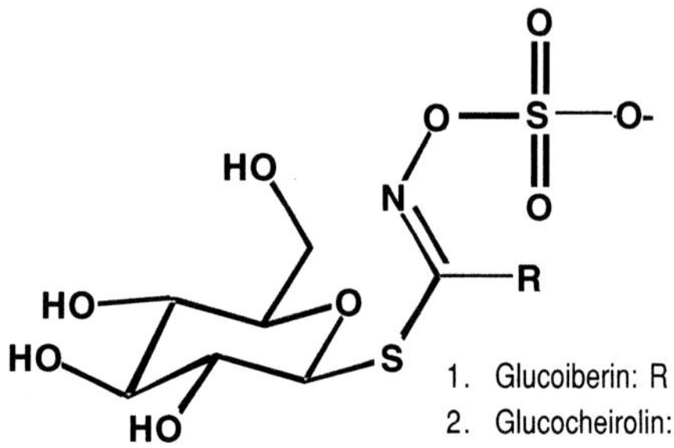

1. Glucoiberin: R = CH_3–SO–$(CH_2)_3$.
2. Glucocheirolin: R = -CH_3-SO_2-$(CH_2)_3$

Fig. 12 Oviposition stimulants (glucosinolates) from the *Erysimum cheiranthoides.*

has shown that high levels of nitrogen result in a dramatic increase in glucosinolate to cardenolide ratios, thus making the plants more acceptable to *P. rapae* for oviposition (Hugentobler and Renwick, 1995). Herbivory on cabbage has been shown to result in greatly increased production of indole glucosinolates, but not aliphatic members of the group (Birch et al. 1990). This might be expected to make these plants more acceptable to *P. rapae*. The invasive cruciferous weed, *Alliaria petiolata*, produces flavonoids that act as oviposition deterrents to *P. napi*, and concentrations of these compounds in the plants vary considerably with season, to provide windows of opportunity at which time the balance is in favor of oviposition (Haribal and Renwick, 2001).

Variable Food Discrimination in *Pieris* Larvae

The diets of lepidopterous larvae are most often determined by the adults, as they choose a plant for oviposition. In general, the larvae of oligophagous and monophagous species are stimulated to feed by specific compounds that are characteristic of their host plants. In the case of *Pieris* species, feeding preferences of the larvae generally follow the behavioral choices of the ovipositing adults. Oviposition stimulants (glucosinolates) also act as feeding stimulants, and oviposition deterrents or related compounds in unacceptable plants are deterrent to the larvae.

However, recent studies have revealed considerable plasticity in larval acceptance of plants as food. Discrimination by *Pieris* larvae was found to depend to a large degree on their previous experience with potential food (Renwick, 2001). Common nasturtium, *Tropaeolum majus*, is an occasional host of *Pieris rapae*. When *P. rapae* eggs are laid on nasturtium, the hatching larvae feed and develop normally. However, larvae that have been reared on cabbage or other crucifers refuse to feed on this plant. Similarly, larvae that have fed on cabbage do not readily accept wheatgerm diet as food, although this medium is generally used for mass rearing of the insect. Subsequent work has shown that the larvae become habituated to deterrents present in nasturtium or in wheatgerm. Larvae reared on a wheatgerm diet appear to remain insensitive to a wide range of deterrents. When these larvae are transferred to a normally unacceptable plant, such as *Erysimum* or *Iberis*, they will feed for some time, even though the compounds that are usually deterrent may be toxic (JAAR unpublished). At the same time, larvae feeding on crucifers develop dependence on, or a type of addiction to glucosinolates that are abundant in the plants. Thus previous dietary experience can profoundly affect the acceptance or rejection of food by these larvae. Electrophysiological experiments have suggested that dietary experience can cause changes in the sensitivity of taste receptors used by the larvae to recognize or assess the suitability of food (van Loon 1990). Thus, plasticity allows the individuals to change their feeding behavior according to availability of food. Without this ability, insects would be likely to die of starvation.

Availability of Food Depends on Environmental Conditions. Plasticity to Change Foods.

Additional studies have shown that seasonal differences in plant chemistry and changes in larval sensitivity to different deterrents may affect the use of introduced plants by endogenous insect populations. Garlic mustard, *Alliaria petiolata*, is an introduced crucifer in North America that is generally avoided by the native *Pieris napi oleracea*. However, some larvae feed on these plants at different times of the year (Chew et al., 1989). We have shown that 6'-O- β-D-glucopyranosyl isovitexin [(6''-O- β-D-glucopyranosyl) -6-C- β-D-glycopyranosyl apigenin] (Fig. 13. 1) and alliarinoside (Fig. 13. 2) present in *A. petiolata* are deterrents to feeding by the larvae (Haribal and Renwick 1998c, Haribal et al. 2001). Interestingly, alliarinoside was deterrent only to the neonate larvae while 6'-O- β-D-glucopyranosyl isovitexin was deterrent to later instars, particularly third and fourth (Renwick et al. 2001). Behavioral observations also suggested that the modes of action of these compounds are different. But the presence of these components in the plants varied as the seasons changed and also in different populations of plants in the fields (Fig.14) (Haribal and Renwick 2001). Thus a window of opportunity exists when the plant chemistry is optimal for insect feeding. Indeed, concentrations of the deterrent alliarinoside and flavonoids in the plants were low when the first brood butterflies were observed in our study area (Fig. 14). Thus variation in larval responses to the balance of negative

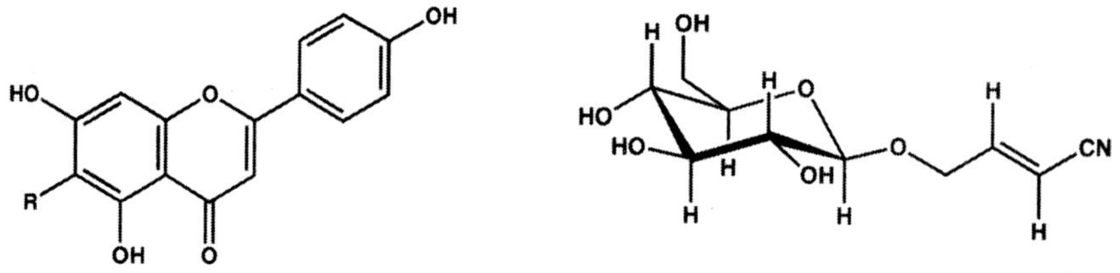

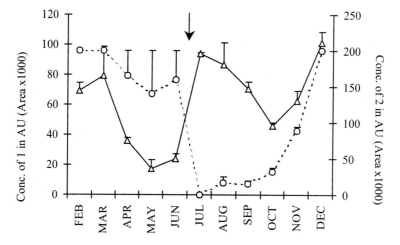

R = 6″ -O-β-D-glucopyranosyl isovitexin
1. 6′-O-β-D-glucopyranosyl isovitexin   2. Alliarinoside

Fig. 13   Feeding deterrents for *Pieris napi oleracea* larvae from *Alliaria petiolata*.

Fig. 14   Seasonal variation in the concentrations of alliarinoside (open circles) and 6-O-b-D-glucopyaranosyl isovitexin (open triangles) in *Alliaria petiolata* collected from Salmon Creek, Ithaca NY. Arrow indicates the window of opportunity where the insects may be able to deal with the plant chemistry.

and positive chemical cues, as well as variation in plant chemistry can play an important role in the ability of *Pieris* species to adapt to an introduced species of plant.

Conclusions

It is clear from these and other examples that the complex interactions occurring between plants and associated organisms involve multidimensional changes in a dynamic chemical world. Plants are remarkably adept at changing their production of chemicals according to their specific needs. In the presence of sunlight, they undergo photosynthesis and synthesize a plethora of chemicals to serve different functions. It is thought that the biosynthetic ability to produce individual groups of chemicals has evolved as a result of multiple stresses, such as pathogen attack, herbivory and environmental changes. These chemicals then serve as defensive agents or otherwise enable the plants to deal with these stresses. But much more rapid changes in chemistry occur as the seasons change, or as the plants are confronted with new ecological challenges. Such changes might be triggered by differences in light levels, day length, precipitation, temperature, or by the presence of competing vegetation.

Herbivores have evolved to use plant chemistry to their own advantage. Phytophagous insects in general depend on chemical information to find suitable food, and numerous studies have shown that insects use particular chemicals as cues to recognize host plants or to avoid non-hosts. But few attempts have been made to show how changes in these plant chemicals within a day or throughout a season can affect the behavior of associated insects.

In the case of zebra swallowtails, which have a very limited range of host plants, a very narrow window of opportunity exists for the butterflies to exploit these plants. As a result of the plants' unique ecology, the butterflies have evolved to recognize different blends of chemicals and for individual butterflies to memorize chemical signals on the basis of previous experience. In addition, some individual swallowtails appear to have evolved the ability to tolerate higher levels of flavonoids. Furthermore, to avoid unsuitable plants later in the season, these insects have evolved a strategy of diapausing.

A similar example is found in *Pieris virginiensis*, which produces only one brood per year, as its host plant, *Dentaria* spp., is very short lived. However, many species of insects have more than one brood in a given season, but have distinct forms for different seasons. In the Satyrinae, dry season and wet season forms are very common. These insects must encounter the seasonal changes in their host plants and, therefore, must have some way of recognizing and adapting to these changes. It appears that different broods may then use different cues in a particular season. Indeed, there is some evidence that zebra swallowtail butterflies in the later season can tolerate higher amounts of flavonoids (MH observations). Although no clear proof of this has been found, the timing of arrival of this brood would indicate that these individuals are physiologically adapted to the changes in chemistry occurring in their hosts.

The migratory life cycle of the monarch butterflies exposes these insects to chemical differences that are based on geographical distribution of their host plants, as well as to the usual environmental stresses. Since different species of *Asclepias* are used as hosts as they move north in spring and early summer, the chemical cues for oviposition change considerably. These butterflies have adapted to use receptors on different appendages to taste the various chemicals that are characteristic of their hosts.

The role of experience in the responses of insects to changing plant chemistry has been emphasized by several studies. In *Pieris* larvae, the sensitivity of receptors involved in perception of chemical cues changes with dietary experience. Similarly, zebra swallowtail butterflies apparently learn to respond to blends of plant volatiles that they experience in the early stages of their host-seeking. As a result, individual butterflies may use different compounds to recognize the same host plants. In a similar way, larvae of the solanaceous specialist, *Manduca sexta*, become dependent on the presence of a specific chemical (indioside D) in potato foliage for continued feeding on alternate foods (del Campo et al. 2001). But the chemistry of different solanaceous plants varies considerably, and we have shown that all host plants of *M. sexta* do not contain indioside D. However, a closely related compound that acts as a feeding stimulant has been identified from *Solanum surattenses* (MH and JAAR, unpublished).

The importance of balancing the input from stimulants and deterrents to assess the acceptability of a plant has been amply demonstrated with the *Pieris* model system. These butterflies have clearly evolved to use different cues to exploit a wide variety of related species. Similar adaptive behavior has been observed in *Papilio* butterfly species. *P. xuthus* and *P. protenor*, have overlapping hosts within the family Rutaceae. Workers in Japan have shown that these species use two different sets of chemical cues to recognize plants for oviposition (Honda 1986, Honda and Hayashi 1995, Nishida 1995). However, little or no information is available whether the plant chemistry varies during the season or in a spatial sense that would allow the insects to use such windows of opportunity to lay eggs. Although the host plants produce a variety of chemicals, these two insect species seem to have overcome the chemical barriers and have evolved to use different cues to recognize the hosts. Some populations of other insects such as flea beetles, which have limited dispersal capability, have evolved to use plants that are generally unsuitable for most of the populations (de Jong and Nielsen 1999, 2002).

We can conclude that phenotypic plasticity in response to variable plant chemistry is evident in all the examples discussed here and insects have overcome changes in plant chemistry by changing physiology, modifying taste receptors, producing different broods, by developing the capacity to recognize multiple chemicals, and from associative learning. Also, within a given population of insects, individuals have different thresholds for different chemical cues. This flexibility in response is clearly responsible for the survival of many insect species in rapidly changing environments. Variation in environmental conditions is particularly dramatic when different geographical locations are compared within temperate or desert regions. In such cases, differences in plant chemistry are expected to be greatest. On the other hand, fewer changes would be expected in the tropics, where weather and amount of daylight are relatively constant throughout the year. Thus insects are likely to exhibit more plasticity in those ecosystems where host chemistry is dramatically changing, as in more temperate or deciduous environments.

Acknowledgements

The studies of zebra swallowtails were funded by NSF Grant # IBN-9420319 and IBN-9986250 to Paul Feeny, Ecology and Evolutionary Biology, Cornell University. It is reported here by MH, with his permission. Work on solanaceous insects was supported by USDA Grant # 2001-35302-09884 to JAAR. We are grateful to Erich Städler for reviewing the early version of this manuscript.

References

Arikawa, K. 2001. Hindsight of butterflies. Bioscience, 51: 219-225.

Arikawa, K., and N. Takagi. 2001. Genital photoreceptors have crucial role in oviposition in Japanese yellow swallowtail butterfly, *Papilio xuthus*. Zoological Science, Tokyo, 18: 175-179.

Baur, R., P. Feeny, and E. Städler. 1993. Oviposition stimulants for the black swallowtail butterfly: Identification of electrophysiologically active compounds in carrot volatiles. Journal of Chemical Ecology, 19: 919-938.

Baur, R., M. Haribal, J. A. A. Renwick, and E. Städler. 1998. Contact chemoreception related to host selection and oviposition behaviour in the monarch butterfly, *Danaus plexippus*. Physiological Entomology, 23: 7-19.

Bernays, E. 1995. Effects of experience on feeding, pp 279-305. *In* R.F. Chapman and, G. de Boer [eds.] 1995. Regulatory mechanisms in insect feeding. Chapman & Hall, New York. USA.

Bernays, E. A., and R. F. Chapman. 1994. Contemporary Topics in Entomology, 2. Host-plant selection by phytophagous insects. 312p. Chapman and Hall, Inc., New York, USA.

Birch, A. N. E., D. W. Griffiths, and W. H. McFarlane Smith. 1990. Changes in forage and oilseed rape *Brassica napus* root rlucosinolates in response to attack by turnip root fly *Delia floralis*. Journal of the Science of Food and Agriculture, 51: 309-320.

Borges, M., P. C. Jepson, and P. E. Howse. 1987. Long-range mate location and close-range courtship behavior of the green stink bug *Nezara viridula* and its mediation by sex pheromones. Entomologia Experimentalis et Applicata, 44: 205-212.

Brower, L. P. 1995. Understanding and misunderstanding the migration of the monarch butterfly (Nymphalidae) in North America: 1857-1995. Journal of the Lepidopterists' Society 49:304-385.

Chew, F. S., Goodwillie, C., and Milburn, N. S. 1989. Larval responses of two *Pieris* butterflies to naturalized Cruciferae. American Zoologist, 29: 167A.

Cushman, J. H., Boggs, C. L.,Weiss, S. B., Murphy, D. D., Harvey, A. W., and Ehrlich, P. R. 1994. Estimating female reproductive success of a threatened butterfly: Influence of emergence time and hostplant phenology. Oecologia-(Berlin), 99: 194-200.

Damman, H., and P. Feeny. 1988. Mechanisms and consequences of selective oviposition by the zebra swallowtail butterfly. Animal Behaviour, 36: 563-573.

de Jong, P. W., and J. K. Nielsen. 1999. Polymorphism in a flea beetle for the ability to use an atypical host plant. Proceedings of the Royal Society Biological Sciences Series B, 266: 103-111.

de Jong, P. W., and J. K. Nielsen. 2002. Host plant use of *Phyllotreta nemorum*: Do coadapted gene complexes play a role? Entomologia Experimentalis et Applicata, 104: 207-215.

del Campo, M. L., C. I. Miles, F. C. Schroeder, C. Müller, R. Booker, and J. A. Renwick. 2001. Host recognition by the tobacco hornworm is mediated by a host plant compound. Nature, 411: 186-189.

Dicke, M., J. Takabayashi, M. A. Posthumus, C. Schuette, and O. E. Krips. 1998. Plant-phytoseiid interactions mediated by herbivore-induced plant volatiles: Variation in production of cues and in responses of predatory mites. Experimental and Applied Acarology, 22: 311-333.

Dicke, M., and van Loop, J. J. A. 2000. Multitrophic effects of herbivore-induced plant volatiles in an evolutionary context. Entomologia Experimentalis et Applicata, 97: 237-249.

Edwards, W. H. 1897. Houghton Mufflin Co. Boston.

Ehrlich, P. R., and P. H. Raven. 1964. Butterflies and plants, a study in coevolution. Evolution, 18: 586-608.

Erbilgin, N., and K. F. Raffa. 2001. Modulation of predator attraction to pheromones of two prey species by stereochemistry of plant volatiles. Oecologia, Berlin, 127: 444-453.

Gouinguené, S. P., and T. C. J. Turlings. 2002. The effects of abiotic factors on induced volatile emissions in corn plants. Plant Physiology, 129: 1296-1307.

Greene, E. 1996. Effects of light quality and larval diet on morph induction in the polymorphic caterpillar *Nemoria arizonaria* (Lepidoptera: Geometridae). Biological Journal of Linnean Scioety. 58:277-285.

Haribal, M., and J. A. A. Renwick. 1998a. Identification and distribution of oviposition stimulants for monarch butterflies in hosts and nonhosts. Journal of Chemical Ecology, 24: 891-904.

Haribal, M., and J. A. A. Renwick. 1998b. Differential postalightment oviposition behavior of monarch butterflies on *Asclepias* species. Journal of Insect Behavior, 11: 507-538.

Haribal, M., and J. A. A. Renwick. 1998c. Isovitexin 6''-*O*-b-D-glucopyranoside: A feeding deterrent to *Pieris napi oleracea* from *Alliaria petiolata*. Phytochemistry, 47: 1237-1240.

Haribal, M., and J. A. A. Renwick. 2001. Seasonal and population variation in flavonoid and alliarinoside content of *Alliaria petiolata*. Journal of Chemical Ecology 27: 1585-1594.

Haribal, M., and P. Feeny. 1998. Oviposition stimulant for the zebra swallowtail butterfly, *Eurytides marcellus*, from the foliage of pawpaw, *Asimina triloba*. Chemoecology, 8: 99-110.

Haribal, M., and P. Feeny. 2003. Combined roles of contact stimulant and deterrents in assessment of host-plant quality by ovipositing Zebra swallowtail butterflies. Journal of Chemical Ecology, 29: 653-670.

Haribal, M., P. Feeny, and C. C. Lester. 1998. A caffeoylcyclohexane-1-carboxylic acid derivative from *Asimina triloba*. Phytochemistry, 49: 103-108.

Haribal, M., Z. Yang, A. B. Attygalle, J. A. A. Renwick, and J. Meinwald. 2001. A cynaoallyl glucoside from *Alliaria petiolata*, as a feeding deterrent for larvae of *Pieris napi oleracea*. Journal of Natural Products, 64: 440-443.

Hazel, W. 1995. The causes and evolution of phenotypic plasticity in pupal colors in swallowtail butterflies, pp 205-210. In M. Scriber, Y. Tsubaki and R. Lederhouse [eds.] 1995. Ecology and evolutionary biology of Papilionidae. J. M. Scientific Publishers, Inc., Gainsville, Florida, USA.

Hazel, W. 2002. The environmental and genetic control of seasonal polyphenism in larval color and its adaptive significance in a swallowtail butterfly. Evolution. 56:342-348.

Hilker, M., and T. Meiners. 2002. Induction of plant responses to oviposition and feeding by herbivorous arthropods: A comparison. Entomologia Experimentalis et Applicata, 104: 181-192.

Honda, K. 1986. Flavanone glycosides as oviposition stimulants in a papilionid butterfly *Papilio protenor*. Journal of Chemical Ecology, 12: 1999-2010.

Honda, K., and N. Hayashi. 1995. Chemical factors in rutaceous plants regulating host selection by two swallowtail butterflies, *Papilio protenor* and *P. xuthus* (Lepidoptera: Papilionidae). Applied Entomology and Zoology. 30: 327-334.

Hugentobler, U., and J. A. A. Renwick. 1995. Effects of plant nutrition on the balance of insect relevant cardenolides and glucosinolates in *Erysimum cheiranthoides*. Oecologia Berlin 102: 95-101.

Huang, X. P., and J. A. A. Renwick. 1997. Feeding deterrents and sensitivity suppressors for *Pieris rapae* larvae in wheat germ diet. Journal of Chemical Ecology, 23: 51-70.

McCall, P. J., T. C. J. Turlings, W. J. Lewis, and J. H. Tumlinson. 1993. Role of plant volatiles in host location by the specialist parasitoid *Microplitis croceipes* Cresson (Braconidae: Hymenoptera). Journal of Insect Behavior. 6: 625-639.

Mewis, I., C. Ulrich, and W. H. Schnitzler. 2002. The role of glucosinolates and their hydrolysis products in oviposition and host-plant finding by cabbage webworm, *Hellula undalis*. Entomologia Experimentalis et Applicata, 105: 129-139.

Nishida, R. 1995. Oviposition stimulants of swallowtail butterflies. *In* R. C. E. Lederhouse [ed.], Swallowtail butterflies: Their ecology and evolutionary biology. Scientific Publishers, Inc., Gainesville, Florida, USA.

Pettersson, E. M., and W. Boland. 2003. Potential parasitoid attractants, volatile composition throughout a bark beetle attack. Chemoecology, 13: 27-37.

Pott, M. B., E. Pichersky, and B. Piechulla. 2002. Evening specific oscillations of scent emission, SAMT enzyme activity, and SAMT mRNA in flowers of *Stephanotis floribunda*. Journal of Plant Physiology, 159: 925-934.

Renwick, J. A. A. 2001. Variable diets and changing taste in plant-insect relationships. Journal of Chemical Ecology, 27: 1063-1076.

Renwick, J. A. A., and X. P. Huang. 1994. Interacting chemical stimuli mediating oviposition by Lepidoptera. Oxford and IBH Publishing, New Delhi, India.

Renwick, J. A. A., C. D. Radke, K. Sachdeva-Gupta, and E. Städler. 1992. Leaf surface chemicals stimulating oviposition by Peiris rapae (Lepidoptera: Pieridae) on cabbage. Chemoecology 3: 33-38.

Renwick, J. A. A., W. Zhang, M. Haribal, A. B. Attygalle, and K. D. Lopez. 2001. Dual chemical barriers protect a plant against different larval stages of an insect. Journal of Chemical Ecology, 27: 1575-1583.

Ruther, J., A. Reinecke, K. Thiemann, T. Tolasch, W. Francke, and M. Hilker. 2000. Mate finding in the forest cockchafer, *Melolontha hippocastani*, mediated by volatiles from plants and females. Physiological Entomology, 25: 172-179.

Sachdev Gupta, K., J. A. A. Renwick, and C. D. Radke. 1990. Isolation and identification of oviposition deterrents to cabbage Butterfly *Pieris rapae* from *Erysimum cheiranthoides*. Journal of Chemical Ecology 16: 1059-1068.

Schoonhoven, L. M., T. Jermy, and J. J. A. van Loon. 1998. Insect-Plant Biology. Chapman and Hall, Cambridge.

Shimoda, T., and M. Dicke. 2000. Attraction of a predator to chemical information related to nonprey: When can it be adaptive? Behavioral Ecology, 11: 606-613.

van Loon, J. J. A., A. Blaakmeer, F. C. Griepink, T. A. van Beek, L. M. Schoonhoven, and A. de Groot. 1992. Leaf surface compound from *Brassica oleracea* (Cruciferae) induces oviposition by *Pieris brassicae* (Lepidoptera: Pieridae). Chemoecology 3: 39-44.

van Loon, J. J. A. 1990. Chemoreception of phenolic acids and flavonoids in larvae of two species of *Pieris*. Journal of Comparative Physiology, A Sensory Neural and Behavioral Physiology, 166: 889-900.

4

Fighting, Flight and Fecundity: Behavioural Determinants of Thysanoptera Structural Diversity

Laurence A. Mound

Scientific Associate, Entomology Department, Natural History Museum, London SW7 5BD; CSIRO Entomology, GPO Box 1700 Canberra, ACT, Australia
Laurence.Mound@csiro.au

Introduction

Many species of Thysanoptera have adults so varied in structure that large and small individuals may not be recognisable as being the same species without collateral biological information. Failure to recognise this intraspecific diversity has resulted in many species being given different names, sometimes in different genera. Some of the many structures involved are illustrated by Ananthakrishnan (1969, 1971), together with the variety of taxa within this insect order that show extraordinary phenotypic plasticity. But although variation is often mentioned in the taxonomic literature, there are few studies on its biological significance, and almost none on the causative agencies. Variation within thrips species thus remains largely the domain of descriptive taxonomists. Its biological significance, and the considerable evolutionary and systematic implications arising from it, merit extensive further study.

Historical Background

Differences in colour and structure between individuals of a species from different populations, or even from the same population, have resulted in many common species being described under several names. During the first half of the 20th century, it was not unusual for taxonomists to apply new scientific names to such structural and colour variants with little or no consideration for their biological or genetic significance. An extreme example of this habit was the description by Bagnall (1916) of two new species, *Thrips flavidus* and *Physothrips flavidus*, based on the same female specimens. At the time Bagnall was writing, these two genera were distinguished solely on whether the antennae had eight or nine segments. Because the specimens he was studying had antennae with either seven or eight segments, Bagnall described two new species in two different genera "to avoid confusion by other workers". The proliferation of names has been particularly excessive for those pest thrips that are widespread geographically and variable in colour and size. For example, about 24 species-group names have been proposed for the Onion Thrips, *Thrips tabaci* Lindeman, and about 14 for the Western Flower Thrips, *Frankliniella occidentalis* (Pergande).

In many fungus-feeding thrips species there is even greater variation in structure between sexes and between large and small males. Such variation resulted in one tropical species, *Ecacanthothrips tibialis* (Ashmead), being given 18 different names (Fig. 1)(Palmer & Mound 1978). Intraspecific variation between the largest and smallest males of *Phoxothrips pugilator* Karny (Fig. 2) resulted in a key to the four subgenera of *Phoxothrips* allocating members of this one species to different subgenera (Haga and Okajima 1975). At times, such confusion arises, not just inadvertently but because of a belief that taxonomy concerns merely the description and naming of entities, and that attempts to understand biology should be left to other scientists. Indeed, descriptive taxonomists have been known to insist on the veracity of their species concepts in the face of objections from field observations that suggest differently. The English entomologist R.S. Bagnall, not only initially placed the males and females of *Idolothrips spectrum* (Haliday) in different genera (1908) but insisted subsequently (1916, 1932) that the Australian entomologist, W.W.Froggatt, must be incorrect in recognising only one variable species. Bagnall, working with a few dead specimens in England, simply failed to appreciate that the Australian Giant Thrips is an abundant insect, forming large colonies widely across Australia and with both sexes varying greatly in body size, facts that were well known to the experienced field biologist Froggatt.

During the second half of the 20th century, taxonomists made considerable advances in recognising the phenotypic plasticity of many thrips species. For example, a review of R. S. Bagnall's collections (Mound 1968) established 134 new synonymies, and a re-examination of some Old World fungus-feeding thrips in nine genera (Palmer & Mound 1978) led to 41 new specific synonymies and 10 new generic synonymies. Such

Fig. 1   *Ecacanthothrips tibialis*, large and small male

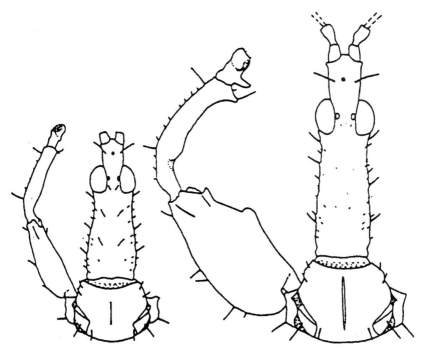

Fig. 2   *Phoxothrips pugilator*, large and small male (Haga and Okajima, 1975)

conclusions owed much to techniques of field observation and collecting devised and taught by Guy D. Morison, who assiduously studied thrips and their host-plants around Aberdeen, Scotland, for about 40 years. These techniques, applied in tropical countries, resulted in extensive population samples from which conclusions on intra- and inter-population variation became possible. Similarly, J. D. Hood at Cornell University, USA, developed very extensive collections of slide-mounted thrips that he had himself collected in Brazil, and these clearly demonstrated that remarkable patterns of within-population structural variation are exhibited by many species.

Earlier descriptive taxonomists in temperate regions had been heavily dependent on haphazardly collected specimens submitted from the tropics by non-specialist collectors. Changes in philosophy and taxonomic practise, including intensive field-work by taxonomists to establish host-plant relationships and create extensive collections based on population samples, led to some remarkable differences in conclusions. D. Moulton in California described about 520 thrips species, many from the Neotropics, and H. Priesner in Austria described about 880 species, many from South East Asia. Largely as a result of the lack of field experience with these faunas, 30% of the species described by both of these authors are now considered synonyms. Similarly, 40% of the 550 species described by Bagnall are now placed into synonymy (Gaston & Mound 1993). In contrast, Hood did most of his own field work and described as new a total of 1065 thrips species. Critical re-examination of Hood's material by subsequent workers has resulted in considerably less than 10% of these being considered to be synonyms.

Proximity of a taxonomist to a fauna under study is, however, no guarantee that intraspecific variation in colour and structure will be assessed satisfactorily. A.A. Girault described 135 species of thrips, mostly from specimens that he collected personally near his home in Brisbane, Australia. More than 50% of Girault's names are now considered synonyms (Mound 1996). Similarly, during the last 10 years the descriptions of many species by workers in Central America have been criticised as not taking into account well-recognised variation in related pest species (Mound & Strassen 2001).

The relationships of intraspecific variation in structure and colour to food quality or quantity and even flower colour, remain largely unstudied. Currently, we can only guess at the physiological stimuli that might lead to the production of winged or wingless adults by some species. The variation in size and structure between large and small males is assumed to be due to food, but whether every individual in a population has the genetic potential to be either large or small remains unknown. Even in pest species, the causes of intraspecific colour differences, also the causes of intraspecific variation in virus vector ability, remain to be studied, whether by breeding experiments or by genetic studies using molecular methods. The concept of phenotypic plasticity in thrips thus remains largely subjective, although potentially it is a highly fertile area for experimentation and biological observation, particularly in tropical countries.

Intraspecific Variation – Genetic and Non-genetic

A complex interaction between various genetic and environmental factors presumably determines the variation that can be observed within many species. Thrips are haplodiploid insects, that is, the males have half the number of chromosomes of the females. Moreover, a natural population rarely comprises more than 30% males (Mound 1992). The remarkable differences in body form that can occur between the sexes in some species of Thysanoptera (Figs 3, 4) are presumably determined genetically. In most Thripidae males are considerably smaller than females, the difference in size being particularly clear in the rare bisexual individuals called gynandromorphs. One such individual, with the left side of the body male but the right side female, was observed walking in circles on a flower surface, because the male side was half as big as the female side (Mound

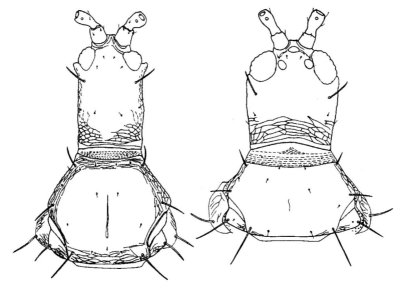

Fig. 3 *Hoplothrips flavicinctus*, wingless male and winged female (Stannard 1968)

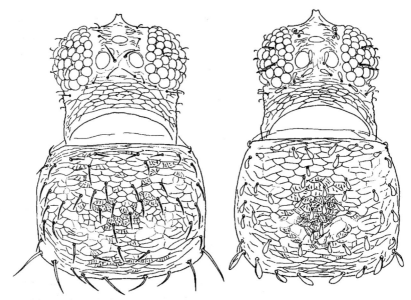

Fig. 4 *Watanabeothrips yasuakii* Okajima, male and female (Okajima 2002).

1971). Male Phlaeothripidae are commonly much larger than females, particularly in fungus-feeding species. Sexual dimorphism in body structure sometimes also involves wing dimorphism. Male Thripidae are commonly flightless and their females winged, such differences presumably being controlled genetically, whereas wing dimorphism in many Phlaeothripidae occurs in both sexes, and is presumably triggered by environmental factors.

Observing structural and colour differences between thrips from different populations can be easy, but recognising the biological significance of such differences is often very difficult. Some species of Thripidae are commonly both smaller and paler when reared at high temperatures than at low temperatures. Indeed, some described species differ only in character states that could well be affected by environmental conditions, and assessing such differences as indicating "new species" in the absence of supporting evidence is particularly facile. Thus *Thrips menyanthidis* Bagnall lives on an aquatic plant in northern Europe in cold boggy places, but it is probably no more than a dark form of the widespread species *Thrips fuscipennis* Haliday (Mound et al. 1976). Similarly, *Aptinothrips nitidulus* Haliday is widespread in salt-marshes in Europe, but is probably an environmentally induced dark form of the ubiquitous grass-living species *Aptinothrips rufus* Haliday (Palmer 1975). There are other pairs of species that may also be environmentally induced variants. For example, *Rubiothrips vitis* Priesner is a minor pest of vines in southern Europe, but it cannot be distinguished from *Rubiothrips sordidus* Uzel, a thrips that lives on *Galium* species, except by its slightly larger size and darker colour (Moritz et al. 2001).

Frankliniella occidentalis, the Western Flower Thrips, is one of the most ubiquitous of thrips pests. One study in California (Bryan & Smith 1956) concluded that this thrips exists in three colour morphs, dark, pale and bicoloured, and that these are determined in a strict Mendelian manner. Despite this genetic conclusion, the authors mentioned that the dark forms were more abundant in early Spring. Subsequent collections in California by the present author indicated that environmental conditions are probably of great importance in determining the colour differences (Fig. 5). In early summer of July 1997, the pale or bicoloured forms of *F. occidentalis* were dominant in the lowlands near Sacramento, but the proportion of dark forms increased with increasing altitude into the Sierre. Moreover, it was observed

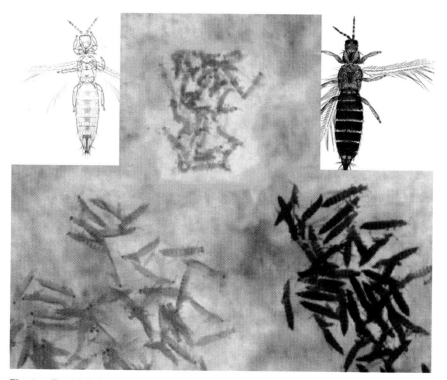

Fig. 5 *Frankliniella occidentalis*, pale males (centre) with yellow and brown females from Californian montane population; inset, a yellow and a brown female.

that dark forms seemed to be particularly dominant in dark coloured flowers. The body colour of this pest thrips has also been noted to vary with the seasons in southern Queensland, Australia, with individuals much darker in frosty winter weather than in high summer. Similarly, in Canberra, the Australian Plague Thrips, *Thrips imaginis* Bagnall, varies greatly in colour, but females from the blue flowers of *Echium* are usually much darker than females from the white flowers of *Sophora*. Critical experimental work on colour variation has not been attempted on either of these abundant insects.

Thrips tabaci Lindeman, the highly polyphagous Onion Thrips that is found worldwide, is also known to vary considerably in size and colour. Individuals that have emerged from over-wintering pupae in Europe are commonly larger and darker than those of the midsummer generations. In Japan, Murai and Toda (2002) reared this species from egg hatching to adult emergence under two temperature regimes – 15ºC and 25ºC. They discovered that the dark colour of adults is determined by low temperatures at the pupal stage, even when the larvae are maintained at high temperatures. In contrast, body size is determined by the temperature during the larval stage, low temperatures giving rise to larger thrips than high temperatures. The difference between the two 'forms' is not merely colour and size, it affects the number and length of setae, as well as the relative lengths of the antennal segments. For example, this species is unusual within the genus *Thrips* in having, with remarkable consistency, four setae rather than three on the distal half of the forewing first vein. However, small individuals often have only three setae, and the largest female studied by the present author had seven setae in this position. Unfortunately, such differences are commonly used by taxonomists to distinguish "species".

The comments in this Section have emphasised how little is known about the normal pattern of variation in common species of Thysanoptera. Such variation may be observed, but the interplay of genetic and environmental factors is rarely studied experimentally. We can thus do little more than guess at the mechanisms controlling such variation. This lack of knowledge can be of considerable economic importance. For example, some populations of *Thrips tabaci* are major vectors of Tospoviruses that severely damage some crops. Other populations do not transmit these viruses (Chatzivassiliou 2002), and currently there is no satisfactory explanation for this inter-population variation.

The following two Sections are concerned with patterns of structural variation that are outside what might reasonably be considered to be "normal variation". These growth patterns in Thysanoptera, sometimes involving extreme allometry, are remarkable, but recent research has begun to provide some understanding of the selection pressures that drive such phenotypic plasticity.

Phenotypic Plasticity in Males – Fight or be Furtive?

Among species of fungus-feeding Thysanoptera males are commonly far more variable in body size than females. This variation occurs not only in highly derived species of both subfamilies of Phlaeothripidae, the Phlaeothripinae and the spore-feeding Idolothripinae, but also in members of basal lineages of the Thysanoptera, the fungus-feeding Merothripidae and some flower-feeding Aeolothripidae. Although such variation in body size is observed most commonly amongst large species, many minute species also show similar variation. Body size variation, moreover, is commonly associated with allometry, and this phenomenon of differential growth rates is expressed in a wide and unpredictable range of body structures.

As indicated above, variation in one fungus-feeding species, *Ecacanthothrips tibialis*, is so great that 18 different species-group names in four different genera have been applied to it by various taxonomists (Fig. 1). This thrips is widespread on dead twigs across the Old World tropics, from East Africa to Australia, apparently feeding on fungal hyphae. The females are effectively monomorphic, although the largest females have rather thick fore femora with a stout median tubercle and a row of tubercles on the fore tibiae, whereas the smallest females have slender fore femora with almost no tubercle and the fore tibiae are almost unarmed. In contrast, the smallest males are similar in appearance to rather small females, but the largest males have the fore femora greatly expanded, almost twice as wide as the head, and the fore coxae long and produced laterally into an acute projection. The disproportionate enlargement of the fore femora and fore coxae with increasing body size in males is a prime example of allometric growth. Some authors have referred to such large males as "oedymerous" and the small males as "gynaecoid", but these terms serve little purpose beyond that of the simple terms, major and minor males, with their behavioural overtones.

The remarkable variation in structure of males in some Phlaeothripinae species remained little more than a taxonomic curiosity until the work on certain northern hemisphere species of *Hoplothrips* by Crespi (1986a, 1988). The males of these fungus hyphal-feeding thrips were shown to compete with each other in bouts of fighting. Crespi claimed that the very largest males usually avoided combat, after first assessing each others' size by standing side-by-side, but that large and medium-sized males indulged in fights that might lead to the death of the vanquished individual. The males of *Hoplothrips pedicularius* Haliday fight to protect their own egg masses, to which various females will contribute after mating with this male. Small males "sneak-mate" whilst their larger brothers are otherwise occupied. Thus, in such species there exists a trade-off in evolutionary selection between food supply and sexual activity; a large body size enhancing male sexual success requires a rich food supply, but a small body size and rapid development is a great advantage when food is less abundant.

These observations and generalisations by Crespi put the variation of body form in Thysanoptera into a new perspective. However, there have been almost no studies on the significance of phenotypic plasticity in the many tropical thrips species that exhibit this phenomenon. *Tiarothrips subramanii* Ramakrishna, a large Oriental spore-feeding thrips, has the length of the preocular projection on the head, also the length of the third antennal segment, proportionally much greater in large individuals than in small ones (Fig. 6), but there are no published observations on the use to which these differences are put by members of this species. The fore tarsal tooth is much larger in the major than in the minor males, and this is probably associated with fighting behaviour. However, the selection pressures that have induced the remarkably elongate third antennal

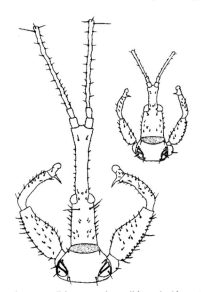

Fig. 6 *Tiarothrips subramannii*, large and small female (Ananthakrishnan, 1969).

segment are not clear, although the undulating form of this segment in major males presumably enhances the strength over that of a simple slender cylinder. Another genus of spore-feeding species, *Mecynothrips*, is mentioned above under its synonymic name, *Phoxothrips*. Members of this genus are found between eastern Africa and the Pacific Islands, but mainly in South East Asia. Large males of some of these species have proportionally much larger fore femora than small males (Fig. 2), with a stout median tubercle (Palmer & Mound 1978). All of the species have an array of stout cheek setae on the head, but large males have more numerous and more robust setae than small males (Fig. 7). Presumably these differences are involved in male/male competitive behaviour, but no recorded observations are available. Ananthakrishnan (1971) provides many further examples of major and minor males.

Elaphrothrips is the largest genus of spore-feeding thrips, with about 150 species found throughout most of the tropics. Colonies of these thrips, sometimes numbering about 100 individuals, can be found on dead twigs

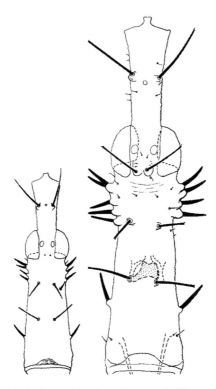

Fig. 7   *Mecynothrips kraussi*, small and large male (Palmer & Mound, 1978).

and particularly within bunches of hanging dead leaves. Large males of many of the species have a remarkable "sickle-shaped" seta curving around the apex of the fore femora, although this seta is not developed in minor males (Fig. 8). The setae on the third and fourth antennal segments are often sexually dimorphic, being long and at right angles to the segment in males but shorter and lying closer to the segment in females. The postocular pair of major setae exhibit a curious example of negative allometry, being very small in major males but longer than the head width in minor males and females. Taxonomic studies on this genus in tropical countries need to be accompanied by careful biological and behavioural observations, because two or more species can sometimes be found at the same site with no obvious resource partitioning between them (Mound & Palmer 1978, Mound & Marullo 1996). Because of the complex patterns of variation in size, structure and sculpture, some of the species distinguished by Johansen (1980, 1983, 1986) may prove to be based on conspecific large and small individuals.

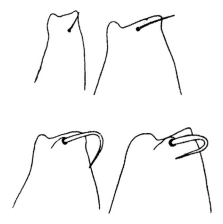

Fig. 8   *Elaphrothrips tuberculatus* Hood, variation in male fore femoral seta (Hood 1955).

Crespi (1986b, 1989) investigated *Elaphrothrips tuberculatus* Hood, a large species that is widespread in eastern North America on dead hanging leaves of *Quercus* trees. Individual males, in contrast to those of *Hoplothrips pedicularius* discussed above, defend single egg-guarding females. These males fight after first assessing their relative body sizes by standing parallel to each other. The abdomen may be flicked at an opponent, but one male will then attempt to climb onto the back of an opponent using the enlarged fore legs to grasp and then stab with the stout fore tarsal teeth. Prolonged fighting bouts occur when males are of similar size, but small males tend to flee in the face of a larger combatant. Similar studies are needed in Central and South America on *Elaphrothrips* species, to understand the complex patterns of structural variation. Members of this genus can be cultured in laboratories on fungal spores on dead leaves, and in tropical countries could provide a rich source of interesting biological, behavioural and developmental studies.

Major and minor males, as indicated above, also occur in some of the basal lineages of Thysanoptera, but whether this indicates a plesiotypic life style or a more recent response to ecological conditions is open to question. *Merothrips floridensis* Watson, living on dead leaves and dead twigs where it presumably feeds on fungal hyphae, is reported from many parts of the world (Mound & O'Neill 1974). Large males of this species have a stout pointed tubercle sub-apically on the fore tibia, but there have been no observations on the significance of this. Similarly, males of *Cycadothrips* species vary greatly in body size, the smallest males being much smaller than females but the largest males much larger (Mound & Terry 2001). The very large males have two pairs of stout thorn-like setae near the abdominal apex, but these setae are slender in the smallest males (Fig. 9). These thrips breed only on the male cones of certain species of *Macrozamia* cycads, a

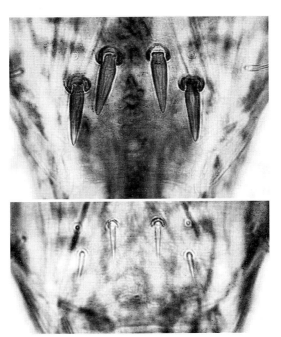

Fig. 9 *Cycadothrips chadwicki,* tergite IX of large and small males.

single cone supporting a population of at least 20000 adults, and presumably there is some form of male competition for females, although this has not been observed.

Even in the family Thripidae there are a few species in which major and minor males are known. Ananthakrishnan (1969) illustrates the forelegs of large and small males of *Perissothrips parviceps* Hood, a species that was subsequently referred by Bhatti (1978) to the genus *Rhamphothrips*. In this genus, the biology of all nine described species remains unknown, but *R. tenuirostris* Karny has been found at more than one site in *Macaranga* flowers (Euphorbiaceae). The major males of a closely related species, taken from *Macaranga gigantea* flowers at Frasers Hill, West Malaysia, have the pronotum and fore coxae greatly elongate, the fore femora enlarged, and the fore tibiae with a stout thumb-like hook at the inner apex. In contrast, minor males are similar to females (Fig. 10).

Phenotypic Plasticity in Females – Fight, or Fly and be Fecund?

Some of the most remarkable intraspecific variation amongst Thysanoptera occurs in females of gall-inducing species in the arid zones of Australia. Several members of the genus *Kladothrips* have monomorphic females, and

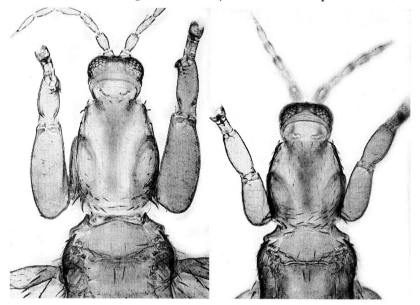

Fig. 10 *Rhamphothrips* ? *tenuirostris*, large and small male.

this is associated with certain other biological attributes. For example, feeding by *K. torus* induces a relatively large gall on *Acacia citrinoviridis*, and within this the founding female lays up to 1000 eggs, each of which hatches and ultimately develops into an adult that is similar in appearance to the foundress. In contrast, several other members of *Kladothrips* have a very different biology with strikingly dimorphic adults of both sexes. A female foundress of *K. habrus* will induce a small gall on *Acacia pendula*, and within this lay up to 10 eggs. These eggs then develop and ultimately produce adults of both sexes that are quite unlike the foundress (Fig. 11), being wingless, with short antennae and massive pronotum and fore legs (Kranz et al. 2001). These wingless adults are termed soldiers, because they act to defend their gall from invading kleptoparasitic thrips species or even microlepidoptera larvae (Crespi 1992). In this species, the wingless adults have a reduced reproductive capacity, further eggs being laid by the original foundress that subsequently develop into fully winged dispersive adults. Such insects with a specialised morph that has reduced reproductive

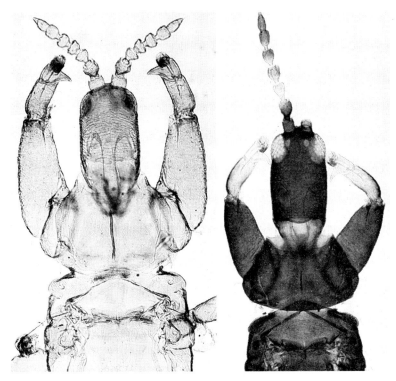

Fig. 11   *Kladothrips kinchega* Wills et al., soldier and foundress females.

capacity are described in behavioural terms as "eusocial" (Crespi & Yanega 1995, Crespi & Choe 1997). A similar life history occurs in several closely related thrips (Wills et al., 2003), although some other members of the same genus produce a single very large and monomorphic brood (Crespi et al., 2004).

The control mechanism, whereby in some *Kladothrips* species the first few eggs produced by a foundress develop into soldiers but her later eggs develop into normal winged adults, remains unknown but it is presumably under the physiological or behavioural control of the foundress. The behaviour of colonial Phlaeothripidae toward their eggs has been little studied, and the presence of adults near an egg mass is usually interpreted as guarding behaviour. However, Moritz (2002) has demonstrated that eggs of *Suocerathrips linguis* Mound & Marullo fail to develop if they are not attended by a group of adults and larvae. In this species it seems that an egg needs to be rolled around by another member of the species if it is to develop satisfactorily. This complex feedback mechanism between adults and developing eggs may also occur in other colonial Phlaeothripidae.

Winged and wingless morphs that are induced environmentally rather than genetically occur in other Thysanoptera. Some of the most startling examples occur in the woody galls found on small branches of trees in the genus *Casuarina* in Australia. The differences between the winged and wingless females of *Phallothrips houstoni* Mound & Crespi are so great that initially the two morphs were assumed to represent different genera (Fig. 12). The head of a winged female is longer than wide whereas that of a wingless female is wider than long, and there are associated differences in the colour and structure of the body (Mound et al. 1998). This species is a kleptoparasite of thrips galls on *Casuarina cristata*, as is the very different looking species *Thaumatothrips froggatti* Karny. Adults of the latter species have long setae on the head and pronotum, but whereas these setae are capitate in winged adults (Fig. 13) they are acute and much longer in micropterous adults.

The woody galls on *Casuarina* species, in which both *P. houstoni* and *T. froggatti* are kleptoparasites, are induced by species of the genus *Iotatubothrips*. Both of the described species in this latter genus are also startlingly dimorphic, winged adults of both sexes having slender antennae and short capitate setae, whereas micropterous adults have shorter antennae and long pointed setae (Fig. 14). These *Casuarina* galls apparently can last for several years, and sometimes contain many hundreds of thrips individuals. As with the *Acacia* thrips discussed above, the flightless individuals apparently defend the gall although in contrast to *Kladothrips*

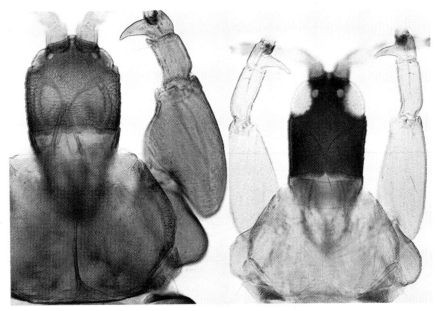

Fig. 12   *Phallothrips houstoni* et al., wingless and winged females.

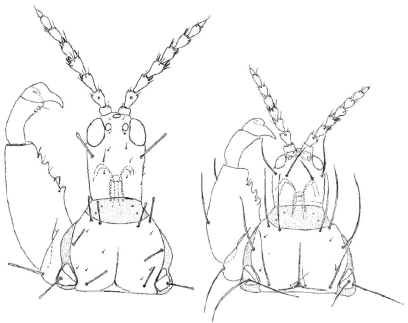

Fig. 13   *Thaumatothrips froggatti*, long winged and short winged females.

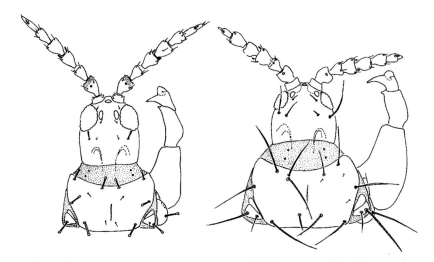

Fig. 14   *Iotatubothrips kranzae*, long winged and short winged females.

soldiers they remain fully reproductive. Ultimately, however, they need to produce dispersive, winged adults. *Iotatubothrips kranzae* on *Casuarina obesa* is particularly remarkable, because the genitalia of winged males are nearly twice as long as the genitalia of wingless males – the only recognised instance of genitalic dimorphism amongst Thysanoptera (Mound et al. 1998).

The remarkable dimorphisms amongst gall thrips in the arid regions of Australia are presumably driven by their selective advantage. It must be very important for these small insects to defend enclosed protective spaces, given the prevailing conditions of high temperature and low humidity. Exposure for quite short periods under these conditions can kill a thrips individual, and moreover new galls can only be induced following the very rare periods of rain (Crespi et al. 2004). Under such circumstances selection pressures have driven the separation into reproductive and dispersive morphs. In contrast, thrips galls that are found in more humid environments may be more easily replaced, with rain and the production of young leaves spread over more months of the year. Certainly, thrips galls in eastern Australia as well as in India and South East Asia often have more species within them. Such galls are referred to by Ananthakrishnan & Raman (1989) as "company galls", but there seem to have been no investigations into the contribution each of the various thrips species makes in the induction of such galls. For example, members of the genus *Euoplothrips* are commonly referred to as "gall thrips", but the range of variation in body form (Fig. 15)

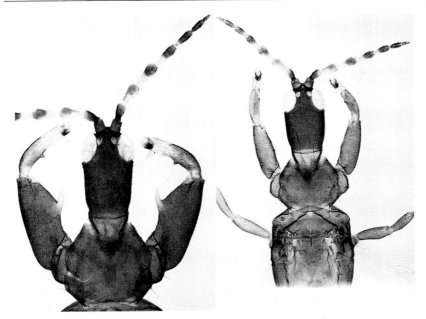

Fig. 15 *Euoplothrips bagnalli*, large and small males.

between large and small individuals of both sexes in *E. bagnalli* Hood suggests that this species indulges in some form of competitive behaviour. In eastern Australia, *E. bagnalli* has been found in rolled-leaf galls of several different Phlaeothripidae and, considering that both sexes are variable, the species is probably a kleptoparasite in these galls rather than a gall initiator. Similarly, *E. platypodae* Marullo has been found in leaf-roll galls of a species of *Gynaikothrips* and this is possibly true of other members of *Euoplothrips* (Marullo 2001).

Although fighting behaviour is particularly obvious in members of the Phlaeothripidae, wing dimorphism occurs in various species in other Thysanoptera families. Amongst the Thripidae, *Frankliniella* is one of the largest genera, with all but 10 of the 160 recognised species always fully winged in both sexes. Most of the 10 exceptions are little known, although one is from sub-Antarctic islands and another from southern South America, and in both these areas wing reduction amongst insects might be expected. However, *F. fusca* Hinds is the tobacco thrips, a well-known pest insect in eastern USA in which both sexes are either fully winged or micropterous. The selection pressures leading to wing dimorphism in this species are not clear, but may be related to the originally fragmented nature of the preferred habitat of the thrips, in small open areas of the extensive fire-prone dense forest of Florida. *F. fusca* is thus essentially like a weedy plant species, building large populations locally when conditions are suitable, but then dispersing.

Two further examples of Thripidae with an irregular and apparently unpredictable alternation between fully winged adults and short-winged adults are *Microcephalothrips abdominalis* Crawford (Fig. 16) in *Helianthus* flowers, and *Sericothrips staphylinus* Haliday on *Ulex* leaves. The former is widespread around the world in areas with warm climate, and its country of origin is not clear. However, the second is from the maritime areas of western Europe, but although now used as a biocontrol agent against its weedy host plant in several countries no pattern has been detected in the production of

Fig. 16 *Microcephalothrips abdominalis*, short winged male with long winged female.

winged forms either in relation to season or plant condition (Hill et al. 2001, Mound et al. 1976). In contrast, variation in wing length in *Thrips nigropilosus* Uzel, a species that does not produce males, has been studied experimentally. Nakao (1999) found that the proportion of short winged adults in a population of this species increased when it was reared at high densities, but that this also resulted in increased mortality and development time.

The development of dispersive and non-dispersive morphs may be an ancestral characteristic of Thysanoptera, since it occurs in the family Merothripidae whose members display the largest number of plesiomorphic character states of this Order. In *Merothrips floridensis* males are always, and females usually, wingless; winged females are produced sporadically and apparently in low numbers. Even in the Aeolothripidae there are a few species that show a similar pattern of wing development. *Aeolothrips albicinctus* Haliday, for example, is a wingless ant mimic that lives as a predator at the base of grasses, but fully winged females are produced occasionally. However, whether this dimorphism is truly plesiotypic, or if it is ecologically driven, remains open to question. The habit may be correlated with the instability of the resource on which a thrips is dependent, whether fungus or arthropod prey. It is much less common amongst thrips species that have adopted seasonally predictable habitats, such as flowers or young leaves.

Environmental Control of Variation

Much of the phenotypic plasticity of thrips species appears to be affected by temperature or food quantity and quality. Body size, and the allometric patterns of growth discussed above, appear to be dependent on the food available to larvae. Similarly, at least amongst fungus-feeding thrips, wing morph seems to be dependent on food availability. When rearing *Hoplothrips fungi* Zetterstedt on *Stereum* fungus on dead wood, the present author found that the thrips colony remained healthy with all adults flightless so long as the *Stereum* remained in good condition. When this fungus was overtaken by an infestation of *Mucor*, probably associated with an increase in relative humidity, the colony produced a large number of winged adults of both sexes. Experimental studies on wing development amongst fungus-feeding thrips could yield interesting information on the factors controlling this dimorphism.

In contrast to environmental or genetic control, wing loss in a few thrips species has a more direct behavioural and physical basis. *Suocerathrips linguis* breeds only on the leaves of *Sanseviera* species, although it does not feed on the plant tissues but on fungi that grow at the bases of the leaves (Moritz 2002). When mating, males grasp the wings (not the body) of a female in their enlarged fore legs, and mating bouts last for many hours. Later, when a mated female cleans her wings these break off at the position where they had been clasped by a male. In colonies of this species it is usual to find that most females have only short wing remnants (Fig. 17) from which the distal two-thirds have been cut (Mound & Marullo 1994). Similarly, mature females in most of the Australian domicile-building thrips of the genus *Dunatothrips* usually have the distal two-thirds of the wings cut away, although the process giving rise to this condition has not been observed (Mound & Morris 2001). Presumably, wing removal in these species is a form of sperm competition, with a male ensuring that a female cannot be inseminated later by another male. The extension of the concept of phenotype to include the behaviour of these thrips enables us to understand why they vary in structure. However, as indicated above, for most thrips

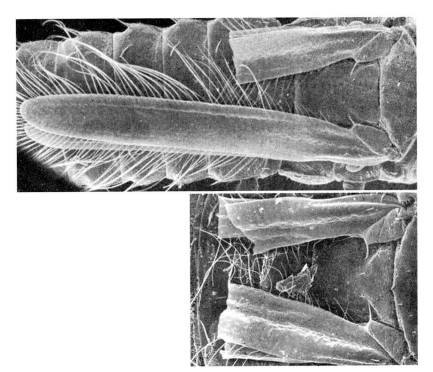

Fig. 17 *Suocerathrips linguis*, females partially and completely de-alate (photo by Gerald Moritz, University of Halle).

species there is no behavioural, developmental or genetic understanding of the wide range of phenotypic plasticity that these insects exhibit.

Acknowledgements

The author is grateful to many colleagues for producing interesting ideas about thrips, particularly Bernie Crespi and David Morris who have developed our ideas on the behavioural significance of variation. Alice Wells criticised an earlier draft of this chapter, and provided much help with field work. Research facilities were provided by CSIRO Entomology at the Australian National Insect Collection, Canberra.

References

Ananthakrishnan, T.N. 1969. Indian Thysanoptera. CSIR Monograph, 1: 1-171.

Ananthakrishnan, T.N. 1971. Trends in intraspecific sex-limited variation in some mycophagous Tubulifera (Thysanoptera). Journal of the Bombay Natural History Society, 67: 481-501.

Ananthakrishnan, T.N. & Raman, A. 1989. Thrips and gall dynamics. E.J.Brill, Leiden. Pp. 120

Bagnall, R.S. 1908. On some new genera and species of Thysanoptera. Transactions of the Natural History Society of Northumberland, 3: 183-217.

Bagnall, R.S. 1916. Brief descriptions of new Thysanoptera. VIII. Annals and Magazine of Natural History, 17: 397-412

Bagnall, R.S. 1932. Brief descriptions of new Thysanoptera. XVII. Annals and Magazine of Natural History, (10) 10: 505-520.

Bhatti, J.S. 1978. Studies in the systematics of *Rhamphothrips*. Oriental Insects, 12: 281-303.

Bryan, D.E. and Smith, R.F. 1956. The *Frankliniella occidentalis* (Pergande) complex in California (Thysanoptera: Thripidae). University of California Publications in Entomology, 10: 359-410.

Chatzivassiliou, E.K. 2002. *Thrips tabaci*: an ambiguous vector of TSWV in perspective. pp. 69-75. *In* Marullo, R. and Mound, L.A. [eds]. Thrips and Tospoviruses: Proceedings of the 7th International Symposium on Thysanoptera. Reggio Calabria.

Crespi, B.J. 1986a. Territoriality and fighting in a colonial thrips, *Hoplothrips pedicularius*, and sexual dimorphism in Thysanoptera. Ecological Entomology, 11: 119-130.

Crespi, B.J. 1986b. Size assessment and alternative fighting tactics in *Elaphrothrips tuberculatus* (Insecta: Thysanoptera). Animal Behaviour, 34: 1324-1335.

Crespi, B.J. 1988. Adaptation, compromise and constraint: the development, morphometrics and behavioral basis of a fighter-flier polymorphism in male *Hoplothrips karnyi*. Behavioral Ecology and Sociobiology, 23: 93-104.

Crespi, B.J. 1989. Sexual selection and assortative mating in subdivided populations of the thrips *Elaphrothrips tuberculatus* (Insecta: Thysanoptera). Ethology, 3: 265-278.

Crespi, B. J. 1992. The behavioral ecology of Australian gall thrips. Journal of Natural History, 26: 769-809.

Crespi, B.J. and Yanega D. 1995. The definition of eusociality. Behavioral Ecology, 6: 109-115.

Crespi, B.J. and Choe J.C. 1997. Evolution and explanation of social systems. Pp 499-524. *in* Crespi B.J. and Choe, J.C. (eds), The Evolution of Social Behavior in Insects and Arachnids. Cambridge University Press.

Crespi, B.J., Morris, D.C. and Mound, L.A. 2004. The Evolution of Ecological and Behavioral Diversity: Australian *Acacia* Thrips as Model Organisms. Australian Biological Resources Study, Canberra. Pp 321

Gaston, K.J. and Mound, L.A. 1993. Taxonomy, hypothesis-testing and the biodiversity crisis. Proceedings of the Royal Society, London, B251: 139-142.

Haga K. and Okajima, S. 1975. Redescription and status of the genus *Phoxothrips* Karny (Thysanoptera). Kontyu, Tokyo, 42: 375-384.

Hill, R.L., Markin, G.P., Gourlay, A.H., Fowler, S.V. and Yoshioka, E. 2001. Host range, release, and establishment of *Sericothrips staphylinus* Haliday (Thysanoptera: Thripidae) as a biological control agent for Gorse, *Ulex europaeus* L. (Fabaceae), in New Zealand and Hawaii. Biological Control, 21: 63-74.

Hood, J.D. 1955. Brasilian Thysanoptera VI. Revista Brasiliera de Entomologia, 4: 51-160.

Johansen, R.M. 1980. El genero *Elaphrothrips* Buffa, 1909 (Thysanoptera: Phlaeothripidae) en Mexico, con nuevas especies y una clave de identificacion. Revista de la Sociedad Mexicana de Historia Natural, 36 [1975]: 195-349.

Johansen, R.M. 1983. El genero *Elaphrothrips* Buffa, 1909 (Thysanoptera: Phlaeothripidae) en el continente Americano; su sistematica, evolucion, biogeografia y biologia. Monog. Inst.Biol.Univ.Auton.Mexico, 1: 1-267.

Johansen, R.M. 1986. Nuevas conceptos taxonomicos y filogeneticos del genero *Elaphrothrips* Buffa, 1909 (Thysanoptera; Phlaeothripidae) del continente americano y descripcion de dos especies nuevas. Anales del Instituto de Biologia. Universidad Nacional de Mexico, 56: 745-868.

Kranz, B. D, Chapman, T. W, Crespi, B. J, & Schwarz, M. P. 2001. Social biology and sex ratios in the gall-inducing thrips, *Oncothrips waterhousei* and *Oncothrips habrus*. Insectes Sociaux, 48: 315-323.

Marullo, R. 2001. Gall thrips of the Austro-Pacific genus *Euoplothrips* Hood (Thysanoptera), with a new species from Australia. Insect Systematics and Evolution, 32: 93-98.

Moritz, G. 2002. The biology of thrips is not the biology of their adults: a developmental view. Pp 259-267. *In* Marullo, R. and Mound, L.A. [eds]. Thrips and Tospoviruses: Proceedings of the 7th International Symposium on Thysanoptera. Reggio Calabria.

Moritz, G., Morris, D.C. and Mound, L.A. 2001. ThripsID – Pest thrips of the world. An interactive identification and information system. Cd-rom published by ACIAR, Australia.

Mound, L.A. 1968. A review of R.S. Bagnall's Thysanoptera collections. Bulletin of the British Museum (Natural History). Ent. Supplement, 11: 1-181.

Mound, L.A. 1971. Sex intergrades in Thysanoptera. Entomologist's Monthly Magazine, 106: 186-189.

Mound, L.A. 1992. Patterns of sexuality in Thysanoptera. In the 1991 Conference on Thrips (Thysanoptera): Insect and Disease considerations in Sugar Maple Management. ed Cameron, E.A. et al. U.S.D.A. Forest Service, General Technical Report NE-161: 2-14.

Mound, L.A. 1996. Thysanoptera, pp 249-336, 397-414 (Index). *In* Wells, A., Zoological Catalogue of Australia. Volume 26. Psocoptera, Phthiraptera, Thysanoptera. Melbourne. CSIRO Australia.

Mound, L.A., Crespi, B.J. and Tucker, A. 1998. Polymorphism and kleptoparasitism in thrips (Thysanoptera: Phlaeothripidae) from woody galls on *Casuarina* trees. Australian Journal of Entomology, 37: 8-16.

Mound LA and Marullo R. 1994. New thrips on mother-in-laws tongue. Entomologist's Monthly Magazine, 130: 95-98.

Mound, L.A. and Marullo, R. 1996. The Thrips of Central and South America: An Introduction. Memoirs on Entomology, International, 6: 1-488.

Mound, L.A., Morison, G.D., Pitkin, B.R. and Palmer, J.M. 1976. Thysanoptera. Handbooks for the Identification of British Insects, 1 (2): 1-79.

Mound, L.A. and Morris, D.C. 2001. Domicile constructing phlaeothripine Thysanoptera from *Acacia* phyllodes in Australia: *Dunatothrips* Moulton and *Sartrithrips* gen.n., with a key to associated genera. Systematic Entomology, 26: 401-419.

Mound, L.A. and O'Neill, K. 1974. Taxonomy of the Merothripidae, with ecological and phylogenetic considerations (Thysanoptera). Journal of Natural History, 8: 481-509.

Mound, L.A. and Terry, I. 2001. Pollination of the central Australian cycad, *Macrozamia macdonnellii*, by a new species of basal clade thrips (Thysanoptera). International Journal of Plant Sciences, 162: 147-154.

Mound, L.A. and zur Strassen, R. 2001. The genus *Scirtothrips* (Thysanoptera: Thripidae) in Mexico: a critique of the review by Johansen & Mojica-Guzmán (1998). Folia Entomologica Mexicana, 40(1): 133-142.

Murai, T. and Toda, S. 2002. Variation of *Thrips tabaci* in colour and size. pp. 377-378. In Maru!lo, R. and Mound, L.A. [eds]. Thrips and Tospoviruses: Proceedings of the 7th International Symposium on Thysanoptera. Reggio Calabria.

Nakao, S. 1999. Brachypterism in thelytokous *Thrips nigropilosus* (Thysanoptera: Thripidae) at high density and its adaptive significance. Entomological Science, 2: 189-194.

Okajima, S. 2002. *Watanabeothrips yasuakii* (Thysanoptera, Thripidae), a new genus and species from Thailand with remarkable sexual dimorphism in setal shape. Special Bulletin of the Japanese Society of Coleopterology, 5: 111-118.

Palmer J.M. 1975. The grass-living genus *Aptinothrips* Haliday (Thysanoptera: Thripidae). Journal of Entomology (B), 44: 175-188

Palmer J.M. and Mound, L.A. 1978. Nine genera of fungus-feeding Phlaeothripidae (Thysanoptera) from the Oriental Region. Bulletin of the British Museum (Natural History). Ent., 37: 153-215.

Stannard, L.J. 1968. The thrips, or Thysanoptera, of Illinois. Bulletin of Illinois Natural History Survey, 29: 215-552.

Wills, T.E., Chapman, T.W., Mound, L.A., Kranz, B.D. and Schwarz, M.P. 2003. The natural history of *Oncothrips kinchega*, a new species of gall-inducing thrips with soldiers (Thysanoptera, Phlaeothripidae). Australian Journal of Entomology, 43: 169-176.

5

Behavioral Diversity and its Apportionment in a Primitively Eusocial Wasp

Raghavendra Gadagkar[1,2] and K. Chandrashekara[1,3*]

[1]*Centre for Ecological Sciences, Indian Institute of Science, Bangalore 560012, India, Email: ragh@ces.iisc.ernet.in*

[2]*Evolutionary and Organismal Biology Unit, Jawaharlal Nehru Centre for Advanced Scientific Research, Jakkur, Bangalore, 560064, India*

[3]*Department of Entomology, University of Agricultural Sciences, GKVK Campus, Bangalore 560065, India*

Phenotypic Plasticity and Behavioural Diversity

Phenotypic plasticity refers to the phenomenon by which the same genotype gives rise to different phenotypes due to environmental variations (Schlichting and Pigliucci 1990). Phenotypic plasticity and its various synonyms have been used with different shades of meaning. West-Eberhard (2003) uses the phrase "developmental plasticity" in the most possible inclusive manner. Phenotypic plasticity can be seen in every aspect of the phenotype, anatomy, morphology, physiology, life history and especially behaviour. Given that the phenotype is a product of the interaction between genes and environment, the existence of phenotypic plasticity is to be expected if the environment varies. That phenotypic plasticity helps organisms to cope with varying environments is also to be expected. The exact mechanisms of causation of phenotypic variation and its role in evolution are however the subject of much discussion (review in Schlichting and Pigliucci 1990, West-Eberhard 2003). In colonies of social insects, phenotypic plasticity takes on an altogether new significance. Because social insect colonies exhibit division of labour, differences between members of the colony are especially useful. While many social insects have

*Present address: Department of Entomology, University of Agricultural Sciences, GKVK Campus, Bangalore, 560065, India.

morphologically differentiated workers specialized in different tasks, in other social insects behavioural differences among morphologically similar individuals help accomplish division of labour. The proximate mechanism that generate behavioural variation among the members of social insect colonies is much debated (reviewed in Bourke and Franks 1995). And yet, there has been insufficient attempt to develop quantitative methods to measure the extent of behavioural variability and understand its distribution among different levels of biological organization in insect societies. Inspired by Lewontin's (1972) attempt to apportion human genetic diversity, here we develop the concept of behavioural diversity and attempt to apportion it between different hierarchical levels within an insect society. We first begin by describing the method used by Lewontin 1972, adapt it to the study of behavioural diversity, apply the adapted methodology to the primitively eusocial wasp *Ropalidia marginata* and discuss the implications of the results. Throughout this chapter we will use the phrase behavioural diversity as a proxy for phenotypic plasticity but of course it is obvious that behavioural diversity is one component of phenotypic plasticity.

The Apportionment of Genetic Diversity in Humans

"It has always been obvious that organisms vary, even to those pre-Darwinian idealists who saw most individual variation as distorted shadows of an ideal. It has been equally apparent, even to those post-Darwinians for whom variation between individuals is the central fact of evolutionary dynamics, that variation is nodal, that individuals fall in clusters in the space of phenotypic description, and that those clusters which we call demes, or races, or species, are the outcomes of an evolutionary process acting on the individual variation. What has changed during the evolution of scientific thought, and is still changing, is our perception of the relative importance of intragroup as opposed to intergroup variation. These changes have been in part a reflection of the uncovering of new biological facts, but only in part. They have also reflected general sociopolitical biases derived from human social experience and carried over into 'scientific' realms." Richard C. Lewontin.

With these characteristic words, Lewontin (1972) introduced his classic paper on the apportionment of human diversity. In this paper, Lewontin invented a new way of measuring and of apportioning genetic diversity in humans. Lewontin's raw data were the then recently acquired measures of isoenzyme frequencies, acquired using starch gel electrophoresis, a method that Lewontin and Hubby, along with Richard Harris had introduced a few years earlier (Harris 1966, Hubby and Lewontin 1966, Lewontin and Hubby

1966). Arguing that a suitable measure of diversity should (1) have a minimum value when there is a single allele, (2) have a maximum value for a given number of alleles when all alleles are in equal frequency, (3) increase as the number of alleles increase and (4) be a convex function of the frequencies of alleles (i.e., the diversity of the sum of two populations should be greater than the mean diversity of the two separate populations), Lewontin chose the Shannon information measure already widely used in ecology to measure species diversity. First proposed by Shannon and Weaver (1949), the Shannon-Weiner index of diversity as it is widely known in the ecological literature, is:

$$H' = -\sum_{i=1}^{S} p_i \ln p_i \tag{1}$$

where p_i is the proportion of the ith species in a collection S species (See also Magurran 1988, Pielou 1975). In Lewontin's calculations, p_i is the proportion if the ith allele in a collection of S alleles at the given locus. Notice that in this method, diversity is calculated separately at each locus for which allelic data are available. Lewontin calculated H' at three levels; at the population level (H_P), using allele frequencies within each population; at the race level (H_R), using allele frequencies averaged over all populations with each race; and at the species level (H_S), using allele frequencies averaged over all populations in all races. Averaging H_P values from all populations to yield $\hat{H}_P$ and averaging over H_R values from all races to yield $\hat{H}_R$, Lewontin apportioned diversity within populations as $\hat{H}_P/H_S$, diversity between populations within races as $(\hat{H}_R - \hat{H}_P)/H_S$ and diversity between races as $(H_S - \hat{H}_R)/\hat{H}_S$. Averaging his results over 17 loci for which allele frequency data were then available for several populations spread over 7 races, Lewontin arrived at the remarkable result that 85.4% of human genetic diversity is located within populations, 8.3% is between populations within races and only 6.3% is between races. This result led Lewontin to the politically satisfying conclusion that *"Human racial classification is of no social value and is positively destructive of social and human relations. Since such racial classification is now seen to be of virtually no genetic or taxonomic significance either, no justification can be offered for its continuance."*

It seems obvious that a similar analysis is needed with respect to other social species as well as with respect to measures other than genetic variability. Here we attempt to fill this need by choosing a very different social species namely the primitively eusocial wasp *Ropalidia marginata* and by focusing on behavioral diversity rather than genetic diversity.

The Insect Societies

Many species of insects exhibit one form or another of social life. To distinguish between species with relatively loose aggregations of individuals from those with more evolved colonial life, a category called eusocial has been defined. Eusocial species are those that exhibit overlap of generations, cooperative brood care and differentiation of colony members into a fertile reproductive caste (queens and kings) and a sterile worker caste. Within the eusocial, it is customary to distinguish two sub-categories, the primitively eusocial which lack morphological differentiation between queens and workers and the advanced or highly eusocial which exhibit morphological caste differentiation (Michener 1969, Wilson, 1971). Ants, bees and wasps in the insect order Hymenoptera and termites in the insect order Isoptera constitute the traditional examples of eusocial species although today the list has expanded to also include some members of Hemiptera, Coleoptera and Thysanoptera among insects, Crustaceans and Arachnida among other arthropods and even a mammal – the naked mole rat (For a review, see Gadagkar 2001).

The insect societies should perhaps be our first choice in extending Lewontin's analysis to other social species. This is because they are potentially more likely to be different from human societies (as compared to vertebrate societies for e.g.,), in the distribution of diversity. Although races are known in some social insect species, race is not the unit of greatest interest here. The social insect colony is an overwhelmingly important unit of organization and hence diversity within and between colonies takes on a level of significance far greater than that of within and between races. In addition, colonies of many social insect species consist of groups of workers specialized in specific tasks and these are called castes. The ergonomic efficiency of social insects and their enormous ecological success is attributed in large part to the presence of such specialized castes within colonies (Oster and Wilson 1978, Wilson, 1990). Thus the distribution of diversity within and between castes is another focus of great interest. Indeed the pattern of distribution of diversity within and between castes is sure to have implications for our understanding of how castes promote ergonomic efficiency and ecological success.

The Primitively Eusocial Wasp *Ropalidia marginata*

Ropalidia marginata (Hymenoptera : Vespidae) is an old world, tropical, primitively eusocial wasp abundantly distributed in peninsular India

(Figure 1) (Gadagkar 2001). Nests are made from paper carton manufactured by the wasps themselves using cellulose fibers from plants in their environment. The nest consists of a number of hexagonal honeycomb like cells in which they rear their brood. The number of adult wasps on a nest range from one to about a hundred although nests with some 20 to 40 individuals are more common than those with more adults than that. New colonies may be founded throughout the year, by one or a group of female wasps. In single foundress nests the lone female builds the nest, lays eggs, forages to feed her larvae and guards her nest from parasites and predators and does everything necessary to bring her offspring to adulthood, unaided by conspecifics. In multiple foundress nests on the other hand, a single female takes control of the colony as its sole reproductive and does little else besides laying eggs. The remaining females function as altruistic sterile workers, building the nest, foraging for food and building material and feeding and caring for the queen's offspring. Eclosing adult male wasps stay in their nest for about a week or so and then leave, never to return, and to lead a nomadic life attempting to mate with foraging females from various nests. Eclosing female wasps on the other hand have several options open to them. One option is to leave their natal nest and initiate a single foundress nest all by themselves. A second option is to leave but join other nests newly initiated by other wasps derived either from their own nests or sometimes

Fig. 1 A typical nest of *Ropalidia marginata* in Bangalore.

from other nests. A third option is to stay back in their natal nest and spend their whole lives as altruistic sterile workers, helping their queen to raise more offspring. Finally, a fourth option is to stay back and work for some time and then, at an opportune moment, drive their queen away and take over their natal nest as its next queen. Without exception, there is one and only one queen in each nest at any given time. Because these wasps live in a tropical environment, their nesting cycles are aseasonal, indeterminate (without a finite duration) and perennial. As queens can be replaced by their younger nestmates, the colonies can be very long-lived.

Behavioral Caste Differentiation

Because of the absence of morphological differences between queens and workers, *R. marginata* is classified as a primitively eusocial species. Although the female wasps in a colony are morphologically identical, they exhibit a great deal of behavioral variability. This variability is however only quantitative. In other words, while most wasps exhibit most behaviors, the rates at which they perform different behaviors and the time they spend in different behaviors varies considerably. In an early study, Gadagkar and Joshi (1983) constructed time-activity budgets for individually identified wasps in two colonies. It is important to note that in doing so, no distinction was made between queens and workers and all behaviors shown by all the wasps were recorded. And yet, most wasps (including queens and workers) spent about 95% of their time in the following six behaviors, namely, sit and groom, sit with raised antennae, sit with raised wings, walk, inspect cells and being absent from the nest. Although all wasps spent about 95% of their time in these six behaviors, the manner in which they allocated their time between these six behaviors varied widely between different wasps (Figure 2). When subjected to principal components analysis and cluster analysis, the wasps were separated into three distinct clusters which were called sitters, fighters and foragers, after their most distinguishing attributes (Figures 3 and 4). An independent examination of the behaviours used in the principal components analysis and other behaviours not used in the principal components analysis shows that sitters spend relatively more time sitting and grooming themselves, fighters show the highest frequencies of dominance behaviours and foragers spend most of their time away from the nest and return with food and building material. Subsequent studies have shown that such a behavioral caste differentiation is a common feature of most *R. marginata* colonies (Chandrashekara and Gadagkar 1991). It should

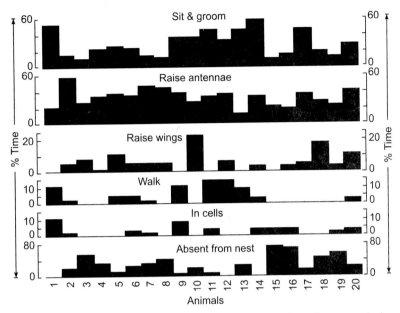

Fig. 2 Time-activity budgets of 20 individually identified wasps drawn from two colonies of *R. marginata*. All wasps spend 85 - 100% (mean ± S.D. = 95.9 ± 0.4) of their time in the six behaviors shown. Note, however, that how the wasps allocate their time among these six behaviors is highly variable. (Redrawn from Gadagkar 1985)

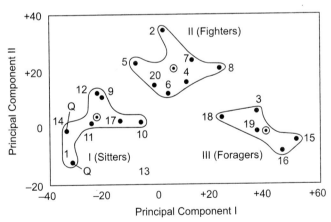

Fig. 3 Behavioural castes of *R. marginata*. Twenty wasps are shown as points in the coordinate space of the amplitudes associated with the first two principal components (based on data in Figure 2). The points fall into three clusters (or castes) by the criterion of nearest centroid. Circled dot = centroid. Q = queen. (Redrawn from Gadagkar and Joshi 1983)

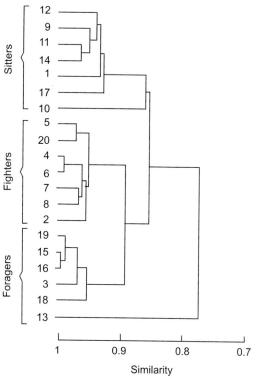

Fig. 4 Hierarchical cluster analysis of the same 20 adults of *R. marginata* that were used in the principal components analysis shown in Figure 3. The similarity between individuals is the Pearson product moment correlation calculated on the basis of the percentage of time spent by the 20 wasps in the same six behaviors used for the principal components analysis (data in Figure 2). A single linkage algorithm was used for clustering. (Redrawn from Gadagkar and Joshi 1983)

be emphasized that this classification of the wasps in a colony into sitters, fighters and foragers was obtained without any *a priori* specification of the behaviors to be used or of the number of clusters to be obtained. Indeed, such behaviors of obvious biological significance as dominance–subordinate behaviors were not used in the classification as they did not rank among the most frequent behaviors which accounted for most of the time of the wasps. Subsequent work has also brought out the biological significance of such behavioral caste differentiation (Gadagkar 2001). Somewhat surprisingly at first sight, the queens always belong to the sitter caste. This suggests that the queens indulge in very few behaviors, conserve and direct their energy towards egg laying. It also suggests that queens do not need to show aggressive dominance behaviors in order to maintain their reproductive monopoly, which they always do with great success. The foragers appear to take on risky extranidal tasks as they probably have the least chance of becoming future queens. This is supported by queen removal experiments where foragers indeed had the lowest probability of becoming replacement queens (Chandrashekara and Gadagkar 1992). Fighters and the remaining sitters seem to share intranidal duties such as building, unloading foragers and brood care and seem to pursue different strategies to maximize their probability of future reproduction. This is supported by queen removal experiments in which both sitters (other than queens) and fighters had approximately equal chances of becoming replacement queens (Chandrashekara and Gadagkar 1992). We have argued elsewhere that such behavioral caste differentiation is the mechanism by which efficient division of labor and relatively harmonious social organization are achieved among a set of competing individuals and that such behavioral caste differentiation may be the forerunner of morphological caste differentiation seen in highly eusocial species (Gadagkar 1991, 2001).

Behavioral Diversity

Following Lewontin's (1972) definition of genetic diversity in human populations, we will here define behavioral diversity of a wasp, in similar fashion. Thus, using the Shannon-Weiner index of diversity (Equation 1) we compute behavioral diversity for every wasp, equating pi as the proportion of time spent by the wasp in the ith behavior. Because different measures of diversity are sometimes used in the ecological literature, in preference to the Shannon-Weiner index, we also repeat our calculation of behavioral diversity using two other commonly used indices of ecological diversity (Magurran 1988). These are the Simpson's index, given by:

$$\lambda = 1 - \sum_{i=1}^{S} p_i^2 \qquad (2)$$

and the so-called N_1, given by:

$$N_1 = \exp(H') \qquad (3)$$

We did not use the other commonly used and highly recommended N_2 (Hill 1973, Peet 1974, Alatalo and Alatalo 1977, Routledge 1979, Ludwig and Reynolds 1988) given by

$$N_2 = 1 / \sum_{i=1}^{S} p_i^2, \qquad (4)$$

because it is not a convex function of the proportion of different species/ behaviors and therefore violates property 4 in Lewontin's list above (Gadagkar 1989). Our data come from 230 h of observations of 12 naturally occurring post-emergence colonies of *R. marginata* in Bangalore, between January 1996 and May 1997. These data have previously been used to show the widespread occurrence of behavioral caste differentiation and to understand its significance (Chandrashekara and Gadagkar 1991). A description of the adult and brood composition of these colonies as also of the numbers of sitters, fighters and foragers in these colonies is given in Table 1.

The Apportionment of Behavioral Diversity

Following Lewontin's method of apportioning human genetic diversity within and between populations and within and between races, we compute behavioral diversity at the individual, caste, colony and species level and apportion behavioral diversity between the following 4 levels:

1. Proportion of diversity within individuals.
2. Proportion of diversity within castes, between individuals.
3. Proportion of diversity within colonies, between castes.
4. Proportion of diversity within species, between colonies.

Before we begin to use data on *Ropalidia marginata*, it is convenient to illustrate that these calculations using a simplified, hypothetical example in Table 2. Our computations of behavioral diversity at different levels and its apportionment between the above mentioned components for our data from 12 *R. marginata* colonies, for all the 3 indices of diversity, are shown in Table 3. The most striking result is that about 72% to 80% of the behavioral diversity is located within individuals, only about 2% to 4% is located within castes between individuals, as much as 14% to 21 % is located within colonies between castes and only about 4% is located within species between colonies. Two major conclusions emerge for these results. The first is that most of the behavioral diversity (about 96%) is within colonies and only about 4% is between colonies. This is very analogous to Lewontin's result that about 93.7% of genetic diversity was located within races and populations and only 6.3% between races. We need to rule out the possibility that our result is because we are using only the 6 common behaviors in which wasps spend 95% of their time and are ignoring the remaining rare behaviors. We therefore repeated these calculations by including, for all wasps, every behavior in which a finite amount of time, however small,

Table 1 Description of the colonies used in this study (Number of animals in the three behavioural castes will not add up to the total number of females in each nest since some animals were not included in the analysis due to inadequate data).

Colony Code	No. of Eggs	No. of Larvae	No. of Pupae	No. of Empty Cells	Total Cells	No. of Males	No. of Females	No. of Sitters	No. of Fighters	No. of Foragers
C01	34	35	25	3	97	0	18	9	6	3
C02	29	33	14	3	79	0	20	6	4	3
C03	16	0	0	34	50	0	12	6	3	1
C04	23	26	7	1	57	0	12	4	2	4
C05	28	20	13	7	68	0	10	2	4	3
C06	75	67	19	30	201	10	39	13	8	10
C07	39	52	41	1	135	0	36	13	9	10
C08	41	58	20	5	124	0	18	8	3	6
C09	62	18	19	3	102	0	8	3	3	1
C10	62	23	22	4	118	0	23	11	6	2
C11	131	37	16	12	196	10	21	13	10	8
C12	61	23	7	21	112	0	16	1	8	6

Table 2   A simplified, hypothetical example illustrating the computation of behavioural diversity

Species	Ropalidia marginata												Mean
Colonies	Colony 1						Colony 2						
Castes	SI		FI		FO		SI		FI		FO		
Individuals	1	2	1	2	1	2	1	2	1	2	1	2	
Beh. 1	0.7	0.6	0.4	0.3	0.1	0.2	0.8	0.7	0.2	0.1	0.2	0.1	
Beh. 2	0.2	0.2	0.4	0.6	0.1	0.2	0.1	0.1	0.5	0.7	0.1	0.3	
Beh. 3	0.1	0.2	0.2	0.1	0.8	0.6	0.1	0.2	0.3	0.2	0.7	0.6	
Div. at Ind. Lev.	0.80	0.95	1.05	0.90	0.64	0.95	0.64	0.80	1.03	0.80	0.80	0.90	0.86
Mean Beh. 1 at Caste Lev.	0.65		0.35		0.15		0.75		0.15		0.15		
Mean Beh. 2 at Caste Lev.	0.20		0.50		0.15		0.10		0.60		0.20		
Mean Beh. 3 at Caste Lev.	0.15		0.15		0.70		0.15		0.25		0.65		
Div. at Caste Lev.	0.90		1.00		0.82		0.73		0.94		0.89		0.88
Mean Beh. 1 at Colony Lev.			0.38						0.35				
Mean Beh. 2 at Colony Lev.			0.28						0.30				
Mean Beh. 3 at Colony Lev.			0.33						0.35				
Div. at Colony Lev.			1.090						1.096				1.093
Mean Beh. 1 at Species Lev.							0.37						
Mean Beh. 2 at Species Lev.							0.29						
Mean Beh. 3 at Species Lev.							0.34						
Div. at Species Lev.							1.094						1.094
Prop. Div. within Ind.							0.86/1.094 = 0.786						
Prop. Div. within Castes, bet. Ind.							(0.88-0.86)/1.094 = 0.018						
Prop. Div. within Col. bet. Castes							(1.093-0.88)/1.094 = 0.195						
Prop. Div. within Species, bet. Colonies							(1.094-1.093)/1.093 = 0.001						

was spent. The results are identical (Table 3) and we can therefore rely on the robustness of these results. The second conclusion concerns the partitioning of behavioural diversity within and between the behavioural castes (see below).

Table 3   Behavioural diversity in the primitively eusocial wasp, *Ropalidia marginata*, at the individual, caste, colony and population levels and it's apportionment between these levels.

Diversity Index	Shannon-Weiner Index		Simpson's Index		N1	
Behaviours used	Common	All	Common	All	Common	All
Mean div. at individual level	0.75	1.03	0.40	0.48	2.20	3.31
Mean div. at caste level	0.80	1.13	0.41	0.49	2.27	3.23
Mean div. at colony level	0.97	1.32	0.53	0.60	2.66	3.79
Mean div. at species level	1.02	1.40	0.56	0.62	2.76	4.04
Prop. div. within individuals	0.74	0.74	0.72	0.77	0.80	0.75
Prop. div. within castes, between individuals	0.04	0.07	0.02	0.02	0.03	0.05
Prop. div. within colonies between castes	0.17	0.14	0.22	0.17	0.14	0.14
Prop. div. within species between colonies	0.04	0.06	0.04	0.04	0.04	0.06

The Robustness of Behavioral Castes

As mentioned above, behavioral caste differentiation into sitters, fighters and foragers was obtained using multivariate statistical analysis of time-activity budget data without any *a priori* decision about which behaviors to use and how many clusters we wish to get. Such an objective, statistical methodology has lead to a classification which seems to have a great deal of biological significance (see above). The results presented here namely that only some 2% to 4 % of behavioral diversity (or up to about 6.6%, if all rare behaviors are considered) is located within castes between individuals and as much as 17 to 21 % (a minimum of 13.7 %, if all rare behaviors are considered) is located between castes within colonies, adds significantly to our confidence in the robustness of this classification. Thus the wasps within a behavioral caste are relatively homogeneous in behavior whereas they vary considerably between castes. To further confirm that this is not a chance occurrence, we have independently re-created the same number of castes, with the same number of individuals in each caste, in each colony. This we did, not by assigning wasps to behavioral castes by principal components analysis of their behavior but by simply drawing random numbers. The results of such a randomization test are shown in Table 4.

Table 4 The apportionment of behavioural diversity within and between castes when castes are determined by randomization.

Diversity Index		Shannon-Weiner Index		Simpson's Index		N1	
Behaviours used		Common	All	Common	All	Common	All
Mean div. at individual level		0.75	1.04	0.40	0.48	2.19	3.02
Mean div. at caste level	Min.	0.92	1.25	0.49	0.57	2.54	3.57
	Max.	0.96	1.29	0.52	0.59	2.63	3.70
Mean div. at colony level		0.97	1.32	0.53	0.60	2.66	3.79
Mean div. at species level		1.02	1.40	0.55	0.62	2.76	4.04
Prop. div. within individuals		0.74	0.74	0.72	0.77	0.80	0.75
Prop. div. within castes, between individuals	Min.	0.17	0.15	0.17	0.14	0.12	0.14
	Max.	0.20	0.18	0.23	0.18	0.16	0.17
Prop. div. within colonies between castes	Min.	0.01	0.02	0.01	0.01	0.01	0.02
	Max.	0.05	0.05	0.06	0.05	0.04	0.05
Prop. div. within species between colonies		0.04	0.06	0.04	0.38	0.04	0.06

Because we created the randomized classification of individuals a thousand times, we provide the maximum and minimum values of the diversity obtained. The important result in Table 4 is that, when castes are obtained randomly, the apportionment of behavioral diversity within and between castes is reversed. Instead of low diversity within castes and high diversity between castes, we now get high diversity within castes (12% - 23%) and low diversity between castes (1% - 6%). This means that our original method of assigning individuals to the sitter, fighter or forager castes, based on its relative behavioral profile, was a robust method indeed, and could not have been obtained by chance alone.

The Social Insect Colony as Noah's ark

In his apportionment of genetic diversity in humans Lewontin found that most of the diversity is within races and populations and very little is between races. He concluded from this that "Racial classification is ... of virtually no ... significance ...". We similarly find that most of the behavioral diversity in the social wasp *R. marginata* is within colonies, castes and individuals and very little diversity exists between colonies (Figure 5). Our conclusion from this is not that classification of the population into colonies is of no significance. On the other hand our important conclusion is that the social insect colony can be thought of a Noah's ark, having within it more or less the entire behavioral diversity of the species. In other words social insect colonies are not aggregations of behaviorally similar individuals but are collections of behaviorally diverse individuals. It is this behavioral diversity that allows social insect colonies to have division of labor, allocate different members of the colony to different tasks in a flexible and regulated manner and achieve ergonomic efficiencies and the resulting ecological successes, unparalleled among solitary species. It is very well known that highly eusocial species with morphological differentiation between reproductives and workers and also sometimes between different worker sub-castes are even more behaviorally diverse. In such species we would expect the distribution of behavioral diversity to be even more skewed in favor of the within colony component. Highly eusocial species are known to employ extraordinary mechanisms to attain high within colony diversity. For example, honeybee queens are known to mate with a large number of different males, gather sperm from these different males and simultaneously generate many different patrilines of daughters. It is also known that the tasks individual bees perform is linked to their patriline of the origin (Page and Robinson 1991, Page et al 1989, 1992, Robinson and Page 1988).

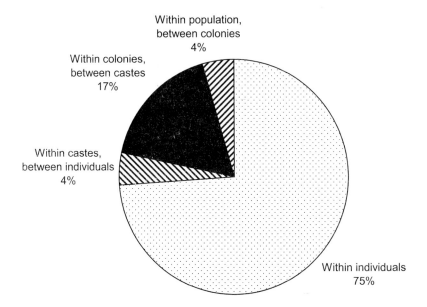

Fig. 5 The apportionment of behavioral diversity, within individuals; within castes, between individuals; within colonies, between castes; and within population, between colonies. This example corresponds to the apportionment of behavioral diversity computed using the common behaviors and the Shannon-Weiner index.

There is growing evidence that the secret of success of that advanced insect societies is their ability to use decentralized, self-organization to achieve complex tasks. In other words the "brain" of an advanced insect society is hardly located in its queen. Instead it is distributed in the tens of thousands of individuals in the colony, as truly distributed intelligence (Detrain et al 1999, Camazine et al 2003). Clearly behavioral diversity is at the heart of the success of social insects.

Acknowledgements

We thank Prof. T.N. Ananthakrishnan for encouragement, Prof. Douglas Whitman for comments on the manuscript and the Department of Science and Technology and the Ministry of Environment and Forests, Government of India for financial support.

References

Alatalo, R. and R. Alatalo. 1977. Components of Diversity: Multivariate Analysis with Interaction. Ecology, 58: 900-906.

Bourke, A.F.G. and N.R. Franks. 1995. Social Evolution in Ants. Princeton University Press, Princeton, New Jersey.

Camazine S., J.-L. Deneubourg, N. R. Franks, J. Sneyd, G. Theraulaz, and E. E. Bonabeau. 2003. Self-organization in Biological Systems. Princeton University Press, Princeton, New Jersey.

Chandrashekara, K. and R. Gadagkar. 1991. Behavioural castes, dominance and division of labour in a primitively eusocial wasp. Ethology, 87: 269-283.

Chandrashekara, K. and R. Gadagkar. 1992. Queen succession in the primitively eusocial tropical wasp *Ropalidia marginata* (Lep.) (Hymenoptera: Vespidae). Journal of Insect Behaviour, 5: 193-209.

Detrain C., J.-L. Deneubourg, and J. M. E. Pasteels. 1999. Information Processing in Social Insects. Birkhäuser Verlag, Basel.

Gadagkar, R. 1985. Evolution of insect sociality : a review of some attempts to test modern theories. Proceedings of the Indian Academy of Sciences (Animal Sciences), 94: 309-324.

Gadagkar, R. 1989. An undesirable property of Hill's diversity index *N2*. Oecologia, 80: 140-141.

Gadagkar R. 1991. *Belonogaster, Mischocyttarus, Parapolybia,* and Independent founding *Ropalidia,* pp 149-190. *In* K. G. Ross and R. W. Matthews. [eds.]. The Social Biology of Wasps. Cornell University Press, Ithaca.

Gadagkar R. 2001. The social biology of *Ropalidia marginata*: Toward understanding the evolution of eusociality. Harvard University Press, Cambridge, Massachusetts.

Gadagkar, R. and N. V. Joshi. 1983. Quantitative ethology of social wasps: time-activity budgets and caste differentiation in *Ropalidia marginata* (Lep.) (Hymenoptera: Vespidae). Animal Behaviour, 31: 26-31.

Harris, H. 1966. Enzyme polymorphisms in man. Proceedings of the Royal Society of London, Series B, 164: 298-310.

Hill, M. O. 1973. Diversity and Evenness: A Unifying Notation and its Consequences. Ecology , 54: 427-432.

Hubby, J. L. and R. C. Lewontin. 1966. A Molecular Approach to the Study of Genic Heterozygosity in Natural Populations. I. The number of alleles at different loci in *Drosophila pseudoobscura.* Genetics, 54: 577-594.

Lewontin, R. C. 1972. The Apportionment of Human Diversity. Evolutionary Biology, 6: 381-398.

Lewontin, R. C. and J. L. Hubby. 1966. A Molecular Approach to the Study of Genic Heterozygosity in Natural Populations. II. Amount of Variation and Degree of Heterozygosity in Natural Populations of *Drosophila pseudoobscura.* Genetics, 54: 595-609.

Ludwig J. A. and J. F. Reynolds. 1988. Statistical Ecology - A Primer on Methods and Computing. John Wiley & Sons, Inc., New York.

Magurran A. E. 1988. Ecological Diversity and Its Measurement. Croom Helm, London.

Michener, C. D. 1969. Comparative social behavior of bees. Annual Review of Entomology, 14: 299-342.

Oster G. F. and E. O. Wilson. 1978. Caste and Ecology in the Social Insects. Princeton University Press, Princeton, New Jersey.

Page, R. E. Jr. and G. E. Robinson. 1991. The Genetics of Division of Labour in Honey Bee Colonies. Advances in Insect Physiology, 23: 117-169.

Page, R. E. Jr., G. E. Robinson, D. S. Britton and M. K. Fondrk. 1992. Genotypic variability for rates of behavioral development in worker honeybees (Apis mellifera L.). Behavioral Ecology, 3: 173-180.

Page, R. E., Jr., G. E. Robinson and M. K. Fondrk. 1989. Genetic specialists, kin recognition and nepotism in honey-bee colonies. Nature, 338: 576-579.

Peet, R. K. 1974. The Measurement of Species Diversity. Annual Review of Ecology and Systematics, 5: 285-307.

Pielou E. C. 1975. Ecological Diversity. John Wiley & Sons, Inc. New York .

Robinson, G. E. and R. E. Jr. Page. 1988. Genetic determination of guarding and undertaking in honey-bee colonies. Nature, 333: 356-358.

Routledge, R. D. 1979. Diversity Indices: Which Ones are Admissible? Journal of Theoretical Biology, 76: 503-515.

Schlichting C. D. and M. Pigliucci. 1990. Phenotypic Evolution - A Reaction Norm Perspective. Sinauer Associates, Inc., Sunderland, Massachusetts.

Shannon C. E. and W. Weaver. 1949. The Mathematical Theory of Communication. University of Illinois Press, Urbana.

West-Eberhard M. J. 2003. Developmental Plasticity and Evolution. Oxford University Press, Oxford.

Wilson E. O. 1971. The Insect Societies. The Belknap Press of Harvard University Press, Cambridge, Massachusetts, USA.

Wilson E. O. 1990. Success and Dominance in Ecosystems: The Case of the Social insects. Ecology Institute, Nordbünte, Germany.

6

Clutch Size Plasticity in the Lepidoptera

James A. Fordyce

Department of Ecology and Evolutionary Biology, University of Tennessee, Knoxville, TN 37996, USA e-mail: jfordyce@utk.edu

Introduction

Insect herbivores are confronted with highly variable food resources. Host plant variation can occur among the potential host plant species that an herbivore might encounter, as well as among and within individuals of the same host plant species (Karban 1992). Plant variation can be a consequence of genetic polymorphisms (Berenbaum and Zangerl 1992), development (West-Eberhard 2003), phenology (Damman 1987; Rausher 1995), or phenotypic plasticity in plant responses to any number of environmental factors, including resource variability and natural enemies, or an interaction among these factors (Karban and Baldwin 1997). Herbivores may cope with this potentially highly variable resource by evolving adaptive plasticity in response to variable host plant related traits (Karban and Agrawal 2002). Traits for which plasticity has been documented include physiological responses to host plant defensive chemistry (Broadway 1996, Feyereisen 1999, Li et al. 2002), behavioral responses to plant defenses (Dussourd 1993), and varying investment in reproductive allocation through clutch size modification (Pilson and Rausher 1988). For example, behavioral plasticity can allow herbivore populations to change host plant preferences based upon the quality and availability of potential host plants (Rausher 1995). These examples of plasticity are believed to be evolutionarily advantageous because they allow herbivores to adjust their phenotype in response to variation in host plant characteristics.

Theory predicts that adaptive plasticity can evolve when environmental variation provides reliable information that an organism can use to respond with the appropriate phenotype or behavior (West-Eberhard 1998, Scheiner

1993, Via et al 1995, Nylin and Gotthard 1998). For example, a population of herbivores might encounter potential host plants that vary across a range of nutritional quality, size, or defensive characteristics. For the herbivore population to evolve adaptive plasticity in response to this host plant variation, the variation encountered in these plant characteristics should provide reliable information on aspects of plant suitability that are relevant to the herbivore. Theory also predicts conditions where canalization of traits will be favored over plasticity. Canalization of traits in invariable environments will be favored by selection, especially when there is a cost associated with exhibiting plasticity (DeWitt et al. 1998, Schlichting and Pigliucci 1998, Pigliucci and Murren 2003). Alternatively, canalization of a compromise phenotype, even for labile traits such as behavior, may be favored in a variable environment when that variation provides no information and plasticity incurs some cost (Scheiner 1993).

With this in mind, certain predictions arise concerning the plasticity of traits believed to be adaptive responses to variation of host plant characteristics. One prediction is that the amount of adaptive plasticity expected in response to host plant variation will vary across a hypothetical spectrum of herbivore host plant specificity. For example, consider the oviposition site choices of a hypothetical ultimate polyphage, the herbivore that selects host plants at random from all available plants. Such an herbivore might show less plasticity and exhibit canalization of traits related to oviposition behavior, essentially exhibiting a compromise strategy that represents the mean optimum phenotype across all of the host plants it potentially encounters. In reality, however, few herbivores truly choose among potential host plants randomly (Jaenike 1990). Rather, insects regarded as polyphagous, or generalists, are usually restricted to using a subset of those plants they potentially encounter as hosts. Constraints on host plant breadth can be imposed by multiple factors, including costs imposed by neural limitations associated with choosing suitable host plants (Janz and Nylin 1997, Bernays 2001, Janz 2003), physiological adaptations to host plant toxins (Nosil 2002, Cornell and Hawkins 2003), plant phenology (Papaj 1986), chemical recognition of suitable host plants (Honda 1995), and behavioral (Dussord and Denno 1994) and morphological (Bernays and Janzen 1988) adaptations to host plant characteristics. Polyphagous insects might evolve adaptive plasticity to the variation encountered among their potential host plants if individuals typically encounter such variability, if such variation provides reliable information about the suitability of the host plant, and if the host plant variability has potential fitness consequences for the herbivore. As we move across the spectrum of herbivore host plant specificity towards monophagy, it is possible that adaptive plasticity would be reduced because of the comparatively lower level of host plant variability encountered by the herbivore population. Selection might favor canalization of locally adapted traits when plasticity incurs some cost, whether it is energetic or an associated risk involved with the lag time to respond to information acquired from the environment, (Schlichting and Pigliucci 1998, Pigliucci and Murren 2003, but see Sultan and Spencer 2002). Thus, we might expect that herbivore populations with multiple available host plant species will exhibit greater adaptive plasticity compared to herbivore populations restricted to one host plant species.

The simplicity of this predictive framework is appealing because it provides expectations of the prevalence of adaptive plasticity based solely on the host plant breadth of the herbivore population. However, given the complexity of factors involved in the interactions between plants and herbivores this model may be too simplistic. The power of predicting adaptive plasticity in this framework will most likely be realized if the relevant aspect of host plant variability encountered represents discontinuous plant traits, such as the presence or absence of particular chemical defenses (Berenbaum and Zangerl 1998) or mechanical defenses such as latex (Dussourd 1993), or phenological aspects of host plants, such as plant ontogeny or senescence (Hayes 1982; Carey 1994). In addition, host plant plasticity should also be considered.

It is now well accepted that plants are not passive participants in their interactions with their herbivores. It has been documented for numerous plant species that they can respond dynamically to herbivore damage through induced responses that include both chemical and structural changes in the plant phenotype (Karban and Baldwin 1997). Herbivore induced plasticity, and the resultant variation encountered by herbivores in host plant quality, should favor the ability of herbivores to reciprocate with counter-plastic responses (Agrawal 2001), or evolve strategies to manipulate or circumvent these responses (Tuomi et al. 1994). In a simplified system, herbivore induced host plant plasticity should not eliminate the trend of reduced adaptive plasticity of herbivores across the feeding spectrum from generalist to specialist herbivores, because a more specialized diet should still result in less variable, more predictable, induced plant responses. Life history plasticity of specialized herbivores may be sacrificed for canalized traits that are selectively favored by the predictable plastic responses of their host plants. Therefore, herbivores may evolve behaviors or life history traits that effectively circumvent or beneficially manipulate herbivore induced plant responses (Rhoades 1985; Karban and Agrawal 2002). This will especially be true for situations where the focal herbivore is the only herbivore attacking the plant, or the herbivore that is most directly affected by the induced plant response. However, most plants are confronted with multiple herbivores that may interact indirectly through induced changes in the host plant phenotype. Through herbivore-induced plasticity, plants can mediate interactions between herbivores of the same or different feeding guilds and herbivores that are spatially and temporally separated (Martinsen 1998, Agrawal and Klein 2000, Denno et al. 2000, Thaler et al. 2001). This added level of complexity might compromise the reliability of the information obtained from a host plant by a herbivore and increase the overall variation in host plant phenotypes, thereby favoring adaptive plasticity over canalization for even highly specialized herbivores.

Empirical studies of herbivore plasticity aimed at testing theoretical predictions are complicated by the fact that the encountered variation can be a consequence of multiple factors. Perhaps more importantly, it is difficult to identify the source of variation that favors plasticity in a particular trait. Plasticity exhibited during one stage of life history may have effects on later stages that are beyond the context of a particular study. In this chapter I will focus on the plasticity of one particular life history trait, clutch size. I will specifically focus on the Lepidoptera, discussing oviposition behavior and egg clustering as it corresponds to the variation encountered in host plant phenotypes. Furthermore, I will discuss how plasticity in clutch size affects later stages of development, emphasizing the importance of simultaneous consideration of clutch size variation and consequences for larval feeding. Finally, I will discuss in this context the adaptive significance of within and among population clutch size variation of the pipevine swallowtail, *Battus philenor* (Papilionidae).

Clutch Size in the Lepidoptera

Clutch size has received considerable attention from researchers because it is fundamentally an important life history trait. Clutch size represents a proportion of the females potential fecundity invested in a particular oviposition event. Clutch size varies widely among and within species of Lepidoptera, with many species laying single eggs, and others laying clutches numbering hundreds of eggs (Stamp 1980). Numerous adaptive and non-adaptive hypotheses have been proposed to explain the evolution of clutch size in the Lepidoptera.

Explanations for large clutches include large female egg load, low availability of suitable host plants, or a combination of these two factors (Tatar 1991, Minkenberg et al. 1992). These hypotheses are essentially non-adaptive from the perspective of the plant-herbivore or herbivore-natural enemy interface, and posit that variation in clutch size is largely a consequence of female motivational state, whereby females lay larger clutches when suitable host plants are encountered infrequently. In these situations, females do not adjust their clutch size in response to host plant quality, competition, natural enemies, or abiotic factors.

Alternatively, adaptive hypotheses have been proposed to explain the significance of large clutch size in the Lepidoptera. These hypotheses can also apply to egg clustering when multiple females oviposit in the same location. Though many species avoid laying eggs near conspecific clutches, possibly to reduce intraspecific competition or density influenced natural enemies, some preferentially add eggs to existing clutches (Ulmer et al. 2003). Some adaptive hypotheses for large clutches and egg clustering propose immediate or direct advantages for the eggs. These advantages include reduced exposure to natural enemies (especially for species that stack or layer eggs), reduced desiccation, and enhanced aposematic display for toxic eggs (Stamp 1980).

However, the majority of hypotheses addressing the adaptive significance of large clutches focus on the consequences realized by larvae, especially early instar larvae where mortality is highest (Stamp 1980; Zalucki et al. 2002). The usual consequence of large clutches is that emerging larvae hatch in synchrony and feed in aggregation for at least a portion of their early development. Feeding in aggregation has received a great deal of attention and a number of hypotheses have been directed at explaining its adaptive significance. For some species, larval aggregation can provide a physiological benefit through thermoregulation (Bryant et al. 2000) and/or maintenance of water balance (Klok and Chown 1999). However, most of the hypotheses regarding the evolution of aggregative feeding pertain to antipredator defense or overcoming host plant defenses.

A defensive function for feeding in aggregation has enjoyed a great deal of support. Though defense can include active defense, such as group thrashing, and indirect defenses, such as mutualism with ants (Axén and Pierce 1998), most work on defense has focused on groups of Lepidoptera that are toxic and warningly colored (Stamp 1980). Here, tight aggregations of larvae may benefit by combining their defenses, or by enhancing their aposematic display, essentially being better advertisers of their unpalatability to potential predators (Evans and Schmidt 1990). Though various phylogenetically based studies have supported these hypotheses (Sillén-Tullberg 1988; Hunter 2000), one important aspect that is often ignored is that aggregation is often most pronounced in the earliest instars, when larvae will likely have less stores of sequestered toxins and often lack the aposematic coloration characteristic of later instars. If defense through combined defenses or enhanced aposematism is important, however, it would be expected that large clutches and aggregative feeding would become canalized and plasticity in clutch size would only be favored if females could accurately predict future predator threats that their larvae will encounter. This is one area of research that deserves more attention.

Numerous hypotheses have been proposed for aggregative feeeding as an adaptation to host plant characteristics. These include aggregation as a strategy for overcoming the physical defenses of the host plant, such as the presence of trichomes (Rathcke and Poole 1975) or the physical toughness of leaves (Clark and Faeth 1997). Alternatively, group feeding can ameliorate plant defenses or enhance plant quality (Denno and Benrey 1997; Fordyce 2003). Feeding in aggregation may also maximize the amount of available food because localized damage allows undamaged regions to continue growing, thus ultimately providing more food for developing larvae (Le Masurier 1994). Large clutch size and aggregative feeding has evolved independently in a number of lineages (Sillén-Tullberg 1988), and it is likely that each of the proposed hypotheses accurately describe the realized benefits of aggregation in some taxa.

Clutch Size Plasticity

Investigations of plasticity in clutch size have largely focused on an ovipositing female's decision to allocate her reproductive investment based upon variation in plant quality encountered among potential host plants. Plant quality and clutch size are expected to co-vary under the assumptions that plant quality predicts offspring growth and survivorship, and that ovipositing females can assess the qualitative variation among potential host plants. Clutch size adjustment of female Lepidoptera in response to host plant variability has been observed in some systems (Table 1).

The examples provided in Table 1 indicate that variation in the amount and quality of food available for larvae, including potential intra-specific competition for that food, is one important factor explaining variation in clutch size. Indeed, only Smits et al. (2001) reported a predictable source of variation that was separate from the amount of foliage potentially available

Table 1   Examples of clutch size plasticity in the Lepidoptera.

Species and Family	Environmental Variable	Response and Consequence	Reference
Halisidota caryae (Arctiidae)	Different plant species	Clutch size varied with plant species	Lawrence (1990)
Paraleucoptera sinuella (Lyonetiidae)	Plant size	Clutch size increases with plant size	Kagata and Ohgushi (2002)
Omphalocera munroei (Pyralidae)	Leaf abundance	Clutch size increased with leaf number; reduces the likelihood that larvae will be required to leave plant	Damman (1991)
Bupalus piniarius (Geometridae)	Plant age	Clutch size increased with leaf (needle) age; eggs on young needles more likely to fall from tree	Smits et al. (2001)
Heliconius hewitsoni (Nymphalidae)	Growth rate of plant shoots	Larger communal clutches on faster growing shoots	Reed (2003)
Mechanitis lysimnia (Nymphalidae)	Plant size; presence of conspecific eggs	Clutch size increased with plant size; clutch size reduced when conspecifics were present	Vasconcellos-Neto et al. (1993)
Battus philenor (Papilionidae)	Plant size	Clutch size increased with plant size; permits larvae to disperse in search of food at a later stage	Pilson and Rausher (1988); Tatar (1991); Fordyce and Nice (2004); * Effect absent in some populations (Fordyce and Nice in 2004)
Luehdorfia japonica (Papilionidae)	Host density	Clutch size increased with plant density	Tsubaki (1995)
Luehdorfia puziloi (Papilionidae)	Leaf abundance	Egg cluster size increased with leaf density	Matsumoto et al. (1993) *Compared multiple populations (same result)
Zerynthia cretica (Papilionidae)	Patch and leaf size	Clutch size increased with plant size	Dennis (1996)

to larvae. They found that female pine loopers lay larger clutches on older pine needles and interpreted this as avoidance of the predictable consequence of growth of young needles, specifically that eggs become detached and consequently hatch on the forest floor, away from edible foliage.

It is interesting to note that although there are numerous hypotheses regarding the adaptive significance of larval aggregative feeding and a plethora of studies testing these hypotheses, studies linking the fitness consequences of variation in clutch size to aggregative feeding are largely absent. For example, feeding in aggregation has been shown to be beneficial to larvae using plants covered with trichomes because aggregations can construct a silk mat, or scaffolding, permitting larvae to circumvent this otherwise effective plant defense (Rathcke and Poole 1975; Fordyce and Agrawal 2001). However, clutch size variation as a response to trichome variation among plants has not been examined. Recently, Ulmer et al. (2003) found that *Mamestra configurata* (Noctuidae) laid larger clutches on leaves with conspecific clutches present, creating large egg clusters that included clutches from multiple females. The consequences for subsequent larval aggregations, however, remain unknown. One study that did investigate a link between clutch size and a consequence of larval aggregation was that of Pierce at al. (1987). They found that myrmecophilous lycaenid larvae benefit when feeding in aggregation because the cost per individual for providing ant rewards is reduced. Females laid fewer clutches on plants with ants excluded compared to plants with ants present. However, they did not investigate whether variation in ant density influenced clutch size, rather they focused on the number of eggs laid and the presence or absence of ants. Linking plasticity in clutch size and the benefits of larval aggregation is one strategy that has been largely ignored for understanding the evolution of this life history trait.

One difficulty for using clutch size plasticity as a tool to understand the evolution of egg clustering and larval aggregation, as discussed previously, is that the factors in the environment that favor plasticity should convey reliable information for the oviposition female. Using clutch size plasticity to test hypotheses that attribute a defensive function for aggregative feeding would require that individual females are able to assess the variation in future threats to larvae based upon information at the time of oviposition. Some studies have found that females will avoid laying eggs on plants where predators are detected, however avoidance is favored over clutch size modification (e.g. Freitas and Oliveira 1996).

One possible strategy for understanding clutch size variation is to explore geographic variation in environmental factors, including host plant variation, clutch size, and clutch size plasticity. Environments usually vary geographically, and we would expect both inter- and intrapopulation responses to that variation. Thus, there may be ecotypic variation where populations from more variable environments should exhibit more plasticity than populations exposed to less variation. For example, if clutch size variation is a consequence of host plant size variation, as in most of the examples in Table 1, we might expect that populations restricted to one host plant species will show less plasticity in clutch size compared to populations with multiple potential host plant species.

Few studies have considered geographic variation in factors leading to clutch size variation. Tsubaki (1995) examined variation in clutch size among populations of a swallowtail butterfly. Here, different populations used different host plant species that varied considerably in morphology. Though the mean clutch size was different among populations using different host plants, clutch size adjustment occurring within and among populations could be explained by variation in the amount of edible foliage available to larvae. Thus, the same type of variation in host plant quality that explained geographic differences in mean clutch size, also explained clutch size plasticity within populations. More studies of among population variation in clutch size and clutch size plasticity are needed to determine if the cues that females use to adjust clutch size are similar throughout a species range.

Clutch Size of the Pipevine Swallowtail: Intra- and Inter-Population Variation

Natural History

The pipevine swallowtail, *Battus philenor*, ranges across most of the southern United States and as far south as Honduras. A disjunct population, designated by some as the distinct sub-species *B. philenor hirsuta*, is restricted to the Central Valley of California. All members of the pipevine swallowtail genus are specialists on plants in the genus *Aristolochia* (Aristolochiaceae) (Racheli and Pariset 1992). Nearly all of the species of *Aristolochia* that have been examined contain toxic alkaloids called aristolochic acids (Chen and Zhu 1987). Pipevine swallowtail larvae sequester these compounds and as a consequence both larvae and adults are chemically protected from many predators (Brower 1958; Fordyce 2000,

2001, Sime 2002). The ecology of this butterfly has received considerable attention from investigators over the past fifty years. The current range of the butterfly, especially the northern limit, can easily be explained by the range limits of its *Aristolochia* host plants, which are largely tropical and subtropical (Pfeifer 1966, 1970). However, there is significant geographic variation in the number and identity of host plants that are available to different populations of the butterfly. Some populations, such as those in the southeastern United States, have multiple *Aristolochia* species available, whereas the California population has only one potential host plant, the endemic *A. californica*. An investigation of mtDNA variation obtained from individuals collected throughout the pipevine swallowtail's range indicated that the greatest genetic diversity is present in the southeastern United States and the lowest genetic diversity is present in California and Mexico. The maximum sequence divergence observed was 1.4%. Though no molecular clock currently exists for the Papilionidae, a generic clock of 2% divergence between lineages per million years (Avise et al. 1987) places the currently observed variation well within the past two million years. Thus, the current range of the pipevine swallowtail is presumably a consequence of recent range expansion from a Pleistocene refuge(s) located in the southeastern United States (Fordyce and Nice 2003). Comparisons of ecological and life history variation among extant populations can be conducted with the knowledge that observed differences likely occurred recently and are not the result of an extended period in isolation. Rather, geographic variation in their life history and behavior may represent recently acquired adaptive traits.

Clutch Size Plasticity in the Pipevine Swallowtail

Mean clutch size of the pipevine swallowtail varies geographically (Spade et al. 1988, Rausher 1995, Fordyce and Nice 2004). Rausher (1995) described an average clutch size of approximately 2.5 eggs/clutch in eastern Texas, with the number of eggs per clutch ranging from 1 to 20. The average clutch size observed in central Texas is 5.0 eggs/clutch, with a range of 1 to 18 (Fordyce and Nice 2004). In California, average clutch size is roughly 13 eggs, ranging from 1 to 86 eggs per clutch (Fig. 1; Fordyce 2003). Spade et al. (1988) described 4 to 6 eggs per plant in Mexico, however these eggs were not laid in the characteristic tight cluster and the authors had no evidence that they represented the clutch of a single female. However, Rausher (1979) observed that females in eastern Texas generally avoid laying eggs on plants where previously laid clutches are present. Larvae feed in aggregation,

Fig. 1 Left: A female pipevine swallowtail, *Battus philenor hirsuta*, on the California pipevine, *A. californica*, prior to oviposition. Right: Egg cluster laid by an individual female on *A. californica*.

though obviously the size of the feeding aggregation differs substantially. The observed variation in the ecology of these populations provides an opportunity to examine the ecological factors that may lead to differentiation in clutch size. Furthermore, it provides the opportunity to investigate the factors that lead to clutch size plasticity and examine how well the theoretical predictions of the evolution of clutch size plasticity hold. Do similar factors observed among populations explain clutch size variation? Are populations that presumably encounter less variation in host plant characteristics (i.e., only one host plant) less plastic in the number of eggs that they lay in a single clutch?

Two factors have been identified that explain some of the variation in clutch size of the pipevine swallowtail; (1) the motivational state of ovipositing females due to egg load (presumably non-adaptive) (Tatar 1991, Pilson and Rausher 1988), and (2) host plant quality, specifically the amount of edible foliage (presumably adaptive) (Pilson and Rausher 1988, Fordyce and Nice 2004). Populations in eastern Texas, which may encounter two potential host plants, *A. serpentaria* and *A. reticulata*, apparently adjust clutch size based upon host plant quality. Specifically, Pilson and Rausher (1988) observed that females discriminated among *A. reticulata* and laid larger clutches on plants that contained more leaves. However, in a population in central Texas where *A. erecta* is the primary host plant, females did not appear to adjust clutch size based upon plant quality, measured as number of leaves, plant height, and patch density (Fordyce and

Nice 2004). Here, however, the average clutch size was slightly larger (5 eggs/clutch) compared to the eastern Texas population (2.5 eggs/clutch). Clutch size variation in the California population, which lays significantly larger clutches than either of the Texas populations studied, was partially explained by host plant characteristics. However, the explanatory variable here was plant height, which is not necessarily a good indication of the amount of edible foliage. Furthermore, plant height explained only 9% of the variation in clutch size (Fordyce and Nice 2004).

The host plants from Texas and California differ greatly in both morphology and growth form. Each of the host plants studied in Texas is a relatively small herbaceous perennial, whereas *A. californica* is a creeping and climbing liana that can densely cover more than 100 m². Based upon this significant size difference, it could be hypothesized that the larger clutches observed in California are a consequence of the California plant being so large and that the same plant size assessment that has been observed in Texas is operating in California. However, this does not appear to be the case. Specifically, analyses of those California plants that were significantly smaller than the Texas plant still received egg clutches that were greater than double the size of that observed for the Texas population (Fordyce and Nice 2004). Thus, it appears that factors other than plant size explain the difference in average clutch size observed between these two populations and that the larger clutches observed in California are not an extension of the same plastic responses to qualitative variation, measured as plant size, observed for the Texas population.

Clutch size is intrinsically linked to the occurrence of aggregative feeding of larvae. Interpopulation differences in average clutch size may correspond to variation in the costs or benefits of aggregative feeding experienced by different populations. Thus, multiple factors can influence the evolution of clutch size traits. Resource availability has been shown to influence clutch size, however other consequences of clutch size acting on the feeding larvae will also influence the evolution of this trait.

Feeding in larger aggregation in California is beneficial because larger groups grow at an increased rate compared to smaller groups (Fordyce 2003). As a consequence, larvae hatching from larger clutches grow faster and spend less time in vulnerable early instars (Fordyce and Agrawal 2001; Fordyce and Shapiro 2003). The effect of increased growth rate is a result of the California host plant's physical characteristics, as well as its plastic responses to larval damage. In California, larger groups are better able to feed on the younger, more nutritious leaves, because they are able to overcome the presence of dense trichomes by communal construction of a silk mat, which permits them to forage unimpeded (Fordyce and Agrawal 2001). Trichomes are not present on the Texas plants that have been studied. The presence of trichomes on the California host plant may be one component favoring larger clutches there. Potentially of more importance is that the damage inflicted by larger groups effectively manipulates the quality of the California host plant (induced suitability), resulting in an accelerated larval growth rate. Specifically, there appears to be a plant mediated indirect interaction among members of a feeding aggregation where plant suitability (measured as the growth rate of larvae) increases as a function of the number of individuals in a feeding aggregation (Fordyce 2003). The advantage of accelerated growth associated with larger feeding aggregations on the California host plant is brief (1 to 3 days) and is only realized by groups inflicting the initial damage. Reciprocal transplant experiments between the California and the *A. erecta* feeding population in central Texas indicated that the effect of larger groups growing faster on *A. californica* is a plant effect, regardless of population origin of larvae. The effect of accelerated growth associated with larger group size was absent on the Texas host plant (Fordyce and Nice 2004). Furthermore, given the relatively smaller size of *A. erecta*, larger feeding aggregations may be detrimental in Texas because larvae are required to disperse in search of a new host plant once the resources of a particular host is exhausted during earlier, more vulnerable stages. Thus, multiple factors, including some that may be unique to particular host plants, are likely imposing selection pressure on optimal clutch size and clutch size plasticity of the pipevine swallowtail.

Resource Variation: Plasticity vs. Canalization

Theory predicts that adaptive plasticity is favored when environmental variation is present, and canalization is favored in invariable environments (Schlichting and Pigliucci 1998, West-Eberhard 2003). Adaptive plastic responses are more likely to occur when environmental cues honestly and reliably predict future environmental conditions (Scheiner 1993). Thus, plasticity in clutch size should be a stable evolutionary strategy when the cues that a female uses to make oviposition decisions vary in a predictable fashion and that canalization of optimal clutch size is favored when little variation occurs. The pipevine swallowtail appears to be subject to both situations. In some populations females apparently modify clutch size based on the amount of available food for larvae. In California, another factor besides food availability may also be an important component explaining clutch size, specifically the ability of larvae to enhance host plant quality as a consequence of feeding in a large aggregation. At least two factors may be influencing the evolution of large clutch size for the California population. First, variation in the amount of edible foliage may favor the retention of some clutch size plasticity in this population. Although the overall plant size in California is large, the majority of the vegetation present is older, lower quality food (Fordyce, unpublished data). Furthermore, it is possible that the large size of the California host plant has imposed selection on the reaction norm of clutch size plasticity itself (Pigliucci and Murren 2003), leading to a larger average clutch size compared to other populations using smaller plants and to the loss of the ability of the California population to "appropriately" adjust clutch size in response to small plants (Fordyce and Nice 2004). Thus, clutch size plasticity in California may not always be adaptive because a female's clutch may exceed the resources available on a given host plant. Second, the beneficial consequences of larvae hatching from larger egg clusters on the California host plant favors increased clutch size. Thus, there may be directional selection in California favoring increased clutch size because of the benefits realized by larger aggregations of larvae. It is not known if clutch size and the realized benefit of feeding in a particular group size co-vary. Future work will need to address whether females in California adjust their clutch size in response to host plant cues that convey reliable information on the realized benefit various groups of larvae will experience.

The California population of the pipevine swallowtail is a recent addition to the California fauna (Fordyce and Nice 2003). The host plant on the other hand may represent a Tertiary relict (Axelrod 1975). *Aristolochia californica* is a long-lived perennial and the pipevine swallowtail is its only specialist herbivore. In fact, herbivores other than the pipevine swallowtail are rarely encountered. It is possible that the California host plant experienced an extremely long period of time without herbivore pressure, relying on the presence of its aristolochic acid toxins to effectively deter generalist herbivores (Chen and Zhu 1987; Park et al. 1997). Upon arriving to California, the pipevine swallowtail may have been introduced to a "naive" host plant that could be exploited simply by modifying clutch size, because increased clutch size and aggregative feeding effectively enhances host plant suitability for the developing larvae.

Geographic variation in the environmental factors that might favor a particular clutch size and/or clutch size plasticity introduces some difficulty when comparing among population variation in this life history trait. Such is the case with the pipevine swallowtail. Pilson and Rausher (1988) found that a combination of plant quality and female motivational state (i.e., egg load) explained some of the clutch size variation in eastern Texas. On the other hand, a similar study by Tatar (1991) found that female egg load best explained clutch size in California, specifically that larger clutches were laid by females with higher egg load. However, when also considering the consequences of clutch size variation for developing larvae on different host plant species, the discrepancy between these two studies may be reconciled. The derived population in California might have larger clutches in response to selection favoring large clutches, whilst retaining some of the clutch size plasticity in response to host plant size that is favored in other populations (Fordyce and Nice 2004). Thus, the larger egg load observed in the California population might be a consequence of selection for a larger clutch size overall. In other words, egg load itself may not be the cause of larger clutches in California, rather selection favoring larger clutches may cause the higher egg load observed in females. Interpreting why the results of the California (Tatar 1991) and Texas (Pilson and Rausher 1988) studies differ demonstrates how different factors can influence clutch size variation observed among populations.

Conclusion

Clutch size plasticity is one strategy for insect herbivores to cope with variable food resources. However, most of the empirical work on clutch size variation has concentrated on one consequence of clutch size, specifically how the size of larval feeding aggregation affects larval survival and growth. Few studies directly examining clutch size plasticity have examined the consequences of clutch size variation beyond the egg stage. Because clutch size and aggregative feeding are usually linked, future studies will benefit by considering both ovipositing female decision-making and larval consequences. Understanding the factors that influence clutch size variation within and among populations may also provide an opportunity to test various hypotheses regarding the evolution of aggregative feeding and phenotypic plasticity. Finally, it behoves investigators to appreciate that populations vary in the factors that favor plasticity vs. canalization towards an optimal clutch size.

Acknowledgements

This manuscript was greatly improved by helpful comments and discussions with Chris Nice, Anurag Agrawal, and Doug Whitman. This work is currently supported by the Department of Ecology and Evolutionary Biology at the University of Tennessee, Knoxville.

References

Agrawal, A. A. 2001. Phenotypic plasticity in the interactions and evolution of species. Science, 294:321-326.

Agrawal, A. A., and C. N. Klein. 2000. What omnivores eat: direct effects of induced plant resistance on herbivores and indirect consequences for diet selection by omnivores. Journal of Animal Ecology, 69:525-535.

Avise, J. C., J. Arnold, R. M. Ball, E. Bermingham, T. Lamb, J. E. Neigel, C. A. Reeb, and N. C. Saunders. 1987. Intraspecific Phylogeography the Mitochondrial Dna Bridge Between Population Genetics and Systematics. Annual Review of Ecology and Systematics 18: 489-522.

Axelrod, D. I. 1975. Evolution and biogeography of madrean-tethyan sclerophyll vegetation. Annals of the Missouri Botanical Garden, 62:280-334.

Axén, A. H., and N. E. Pierce. 1998. Aggregation as a cost-reducing strategy for lycaenid larvae. Behavioral Ecology, 9:109-115.

Berenbaum, M.R. and Zangerl, A.R. 1992. Genetics of secondary metabolism and herbivore resistance in plants. II. Ecological and evolutionary processes. In *Herbivores: Their Interactions with Secondary Plant Metabolites* (Eds., G.A. Rosenthal & M.R. Berenbaum), pp. 415-438. Academic Press, San Diego

Berenbaum, M. R., and A. R. Zangerl. 1998. Chemical phenotype matching between a plant and its insect herbivore. Proceedings of the National Academy of Sciences of the United States of America, 95:13743-13748.

Bernays, E. A. 2001. Neural limitations in phytophagous insects: implications for diet breadth and evolution of host affiliation. Annual Review of Entomology, 46:703-727.

Bernays, E. A., and D. H. Janzen. 1988. Saturniid and sphingid caterpillars: two ways to eat leaves. Ecology, 69:1153-1160.

Broadway, R. M. 1996. Dietary proteinase inhibitors alter complement of midgut proteases. Archives of Insect Biochemistry and Physiology, 32:39-53.

Brower, J. V. Z. 1958. Experimental studies of mimicry in some North American butterflies. Part II. *Battus philenor* and *Papilio troilus*, *P. polyxenes* and *P. glaucus*. Evolution, 12:123-136.

Bryant, S. R., C. D. Thomas, nd J. S. Bale. 2000. Thermal ecology of gregarious and solitary nettle-feeding nymphalid butterfly larvae. Oecologia, 122:1-10.

Carey, D. B. 1994. Diapause and the host plant affiliations of lycaenid butterflies. Oikos, 69:259-266.

Chen, Z.-L., and D.-Y. Zhu. 1987. Aristolochia Alkaloids. Pp. 29-65 *in* A. Brossi, ed. The Alkaloids: Chemistry and Pharmacology. Academic Press, Inc., San Diego.

Clark, B. R., and S. H. Faeth. 1997. The consequences of larval aggregation in the butterfly *Chlosyne lacinia*. Ecological Entomology, 22:408-415.

Cornell, H. V., and B. A. Hawkins. 2003. Herbivore responses to plant secondary compounds: A test of phytochemical coevolution theory. American Naturalist, 161:507-522.

Damman, H. 1987. Leaf quality and enemy avoidance by the larvae of a pyralid moth. Ecology, 68:88-97.

Damman, H. 1991. Oviposition behavior and clutch size in a group-feeding pyralid moth, Omphalocera munroei. Journal of Animal Ecology, 60:193-204.

Dennis, R. L. H. 1995. Oviposition in Zerynthia cretica (Rebel, 1904): Loading on leaves, shoots and plant patches (Lepidoptera, Papilionidae). Nota Lepidopterologica, 18:3-15.

Denno, R. F., and B. Benrey. 1997. Aggregation facilitates larval growth in the neotropical nymphalid butterfly Chlosyne janais. Ecological Entomology, 22:133-141.

Denno, R. F., M. A. Peterson, C. Gratton, J. Cheng, G. A. Langellotto, A. F. Huberty, and D. L. Finke. 2000. Feeding-induced changes in plant quality mediate interspecific competition between sap-feeding herbivores. Ecology, 81:1814-1827.

DeWitt, T. J., A. Sih, and D. S. Wilson. 1998. Costs and limits of phenotypic plasticity. Trends in Ecology & Evolution, 13:77-81.

Dussourd, D. E. 1993. Foraging with finesse: caterpillar adaptations for circumventing plant defense. Pp. 92-131 in N. E. Stamp and T. M. Casey, eds. Caterpillars: ecological and evolutionary constraints on foraging. Chapman and Hall, London.

Dussourd, D. E., and R. F. Denno. 1994. Host-Range of Generalist Caterpillars - Trenching Permits Feeding on Plants with Secretory Canals. Ecology, 75:69-78.

Evans, D.L., and J.O. Schmidt (eds.). 1990. Insect Defenses. Suny Press, Albany, N.Y. 482 pp.

Feyereisen, R. 1999. Insect P450 enzymes. Annual Review of Entomology, 44:507-533.

Fordyce, J. A. 2000. A model without a mimic: Aristolochic acids from the California pipevine swallowtail, Battus philenor hirsuta, and its host plant, Aristolochia californica. Journal of Chemical Ecology, 26:2567-2578.

Fordyce, J. A. 2001. The lethal plant defense paradox remains: inducible host-plant aristolochic acids and the growth and defense of the pipevine swallowtail. Entomologia Experimentalis Et Applicata, 100:339-346.

Fordyce, J. A. 2003. Aggregative feeding of pipevine swallowtail larvae enhances hostplant suitability. Oecologia, 135:250-257.

Fordyce, J. A., and A. A. Agrawal. 2001. The role of plant trichomes and caterpillar group size on growth and defence of the pipevine swallowtail Battus philenor. Journal of Animal Ecology, 70:997-1005.

Fordyce, J. A., and C. C. Nice. 2003. Contemporary patterns in a historical context: Phylogeographic history of the pipevine swallowtail, Battus philenor (Papilionidae). Evolution, 57:1089-1099.

Fordyce, J. A., and C. C. Nice. 2004. Geographic variation in clutch size and a realized benefit of aggregative feeding. Evolution, 58:447-450.

Fordyce, J. A., and A. M. Shapiro. 2003. Another perspective on the slow-growth/high-mortality hypothesis: Chilling effects on swallowtail larvae. Ecology, 84:263-268.

Freitas, A. V. L., and P. S. Oliveira. 1996. Ants as selective agents on herbivore biology: Effects on the behaviour of a non-myrmecophilous butterfly. Journal of Animal Ecology, 65:205-210.

Hayes, J. L. 1982. A study of the relationships of diapause phenomena and other life history characters in temperate butterflies. The American Naturalist, 120:160-170.

Honda, K. 1995. Chemical basis of differential oviposition by lepidopterous insects. Archives of Insect Biochemistry and Physiology, 30:1-23.

Hunter, A. F. 2000. Gregariousness and repellent defences in the survival of phytophagous insects. Oikos, 91:213-224.

Jaenike, J. 1990. Host Specialization in Phytophagous Insects. Annual Review of Ecology and Systematics, 21:243-274.

Janz, N. 2003. The cost of polyphagy: oviposition decision time vs error rate in a butterfly. Oikos, 100:493-496.

Janz, N., and Nylin, S. 1997. The role of female search behaviour in determining host plant range in plant feeding insects: a test of the information processing hypothesis. Proceedings of the Royal Society, Series B, 264:701-707.

Kagata, H., and T. Ohgushi. 2002. Clutch size adjustment of a leaf-mining moth (Lyonetiidae: Lepidoptera) in response to resource availability. Annals of the Entomological Society of America, 95:213-217.

Karban, R. 1992. Plant Variation: Its effects on populations of herbivorous insects. Pp. 195-215 *in* R. S. Fritz and E. L. Simms, eds. Plant resistance to herbivores and pathogens. University of Chicago Press, Chicago.

Karban, R., and A. A. Agrawal. 2002. Herbivore offense. Annual Review of Ecology and Systematics, 33:641-664.

Karban, R., and I. T. Baldwin. 1997. Induced responses to herbivory. University of Chicago Press, Chicago.

Klok, C. J., and S. L. Chown. 1999. Assessing the benefits of aggregation: Thermal biology and water relations of anomalous Emperor Moth caterpillars. Functional Ecology, 13:417-427.

Lawrence, W. S. 1990. The effects of group size and host species on development and survivorship of a gregarious caterpillar *Halisidota caryae* (Lepidoptera: Arctiidae). Ecological Entomology, 15:53-62.

Le Masurier, A. D. 1994. Costs and benefits of egg clustering in *Pieris brassicae*. Journal of Animal Ecology, 63:677-685.

Li, X., M. A. Schuler, and M. R. Berenbaum. 2002. Jasmonate and salicylate induce expression of herbivore cytochrome P450 genes. Nature, 419:712-715.

Martinsen, G. D., E. M. Driebe, and T. G. Whitham. 1998. Indirect interactions mediated by changing plant chemistry: Beaver browsing benefits beetles. Ecology, 79:192-200.

Matsumoto, K., F. Ito, and Y. Tsubaki. 1993. Egg cluster size variation in relation to the larval food abundance in *Luehdorfia puziloi* (Lepidoptera: Papilionidae). Researches on Population Ecology, 35:325-333.

Minkenberg, O. P. J. M., M. Tatar, and J. A. Rosenheim. 1992. Egg Load as a Major Source of Variability in Insect Foraging and Oviposition Behavior. Oikos, 65:134-142.

Nosil, P. 2002. Transition rates between specialization and generalization in phytophagous insects. Evolution, 56:1701-1706.

Nylin, S., and K. Gotthard. 1998. Plasticity in life-history traits. Annual Review of Entomology, 43:63-83.

Papaj, D. R. 1986. Shifts in foraging behavior by a Battus philenor population: Field evidence for switching by individual butterflies. Behavioral Ecology and Sociobiology, 19:31-40.

Park, S.-J., S.-G. Lee, S.-C. Shin, B.-Y. Lee, and Y.-J. Ahn. 1997. Larvicidal and antifeeding activities of Oriental medicinal plant extracts against four species of forest insect pests. Applied Entomology and Zoology, 32:601-608.

Pfeifer, H. W. 1966. Revision of the North and Central American hexandrous species of *Aristolochia* (Aristolochiaceae). Annals of the Missouri Botanical Garden, 53:115-196.

Pfeifer, H. W. 1970. A taxonomic revision of the pentandrous species of *Aristolochia*. The University of Connecticut Publication Series:134.

Pierce, N. E., R. L. Kitching, R. C. Buckley, M. F. J. Taylor, and K. F. Benbow. 1987. The Costs and Benefits of Cooperation Between the Australian Lycaenid Butterfly *Jalmenus-Evagoras* and Its Attendant Ants. Behavioral Ecology and Sociobiology, 21:237-248.

Pigliucci, M., and C. J. Murren. 2003. Perspective: Genetic assimilation and a possible evolutionary paradox: Can macroevolution sometimes be so fast as to pass us by? Evolution, 57:1455-1464.

Pilson, D., and M. D. Rausher. 1988. Clutch Size Adjustment by A Swallowtail Butterfly. Nature, 333:361-363.

Racheli, T., and L. Pariset. 1992. Il genere *Battus* tassonomia e storia naturale. Fragmenta Entomologica (Roma) (Supplemento), 23:1-163.

Rathcke, B. J., and R. W. Poole. 1975. Coevolutionary race continues: butterfly larval adaptation to plant trichomes. Science, 187:175-176.

Rausher, M.D. 1979. Egg recognition: its advantage to a butterfly. Animal Behavior, 27:1034-1040.

Rausher, M. D. 1995. Behavioral ecology of oviposition in the pipevine swallowtail, *Battus philenor*. Pp. 53-62. *in* J. M. Scriber, Y. Tsubaki and R. C. Lederhouse, eds. Swallowtail butterflies: Their ecology and evolutionary biology. Scientific Publishers, Gainesville, FL.

Reed, R.D. 2003. Gregarious oviposition and clutch size adjustment by a *Heliconius* butterfly. Biotropica, 35:555-559.

Rhoades, D. F. 1985. Offensive-defensive interactions between herbivores and plants: Their relevance in herbivore population dynamics and ecological theory. American Naturalist, 125:205-238.

Scheiner, S. M. 1993. Genetics and evolution of phenotypic plasticity. Annual Review of Ecology and Systematics, 24:35-68.

Schlichting, C.D., and Pigliucci, M. 1998. Phenotypic Evolution: a reaction norm perspective. Sinauer Associates, Sunderland, MA. 387 pp.

Sillen-Tullberg, B. 1988. Evolution of gregariousness in aposematic butterfly larvae: A phylogenetic analysis. Evolution, 42:293-305.

Sime, K. 2002. Chemical defence of *Battus philenor* larvae against attack by the parasitoid *Trogus pennator*. Ecological Entomology, 27:337-345.

Smits, A., S. Larsson, and R. Hopkins. 2001. Reduced realised fecundity in the pine looper *Bupalus piniarius* caused by host plant defoliation. Ecological Entomology, 26:417-424.

Spade, P., H. Tyler, and J. W. Brown. 1988. The biology of seven troidine swallowtail butterflies (Papilionidae) in Colima, Mexico. Journal of Research on the Lepidoptera, 26:13-26.

Stamp, N. E. 1980. Egg deposition patterns in butterflies: Why do some species cluster their eggs rather than deposit them singly? The American Naturalist, 115:367-380.

Sultan, S. E., and H. G. Spencer. 2002. Metapopulation structure favors plasticity over local adaptation. American Naturalist, 160:271-283.

Tatar, M. 1991. Clutch size in the swallowtail butterfly, *Battus philenor*: The role of host quality and egg load within and among seasonal flights in California [USA]. Behavioral Ecology and Sociobiology, 28:337-344.

Thaler, J. S., M. J. Stout, R. Karban, and S. S. Duffey. 2001. Jasmonate-mediated induced plant resistance affects a community of herbivores. Ecological Entomology, 26:312-324.

Tsubaki, Y. 1995. Clutch size adjustment by *Luehdorfia japonica*. Pp. 63-70 *in* J. M. Scriber, Y. Tsubaki and R. C. Lederhouse, eds. Swallowtail butterflies: Their ecology and evolutionary biology. Scientific Publishers, Scientific Publishers, Gainesville, FL.

Tuomi, J., E. Haukioja, T. Honkanen, and M. Augner. 1994. Potential benefits of herbivore behaviour inducing amelioration of food-plant quality. Oikos, 70:161-166.

Ulmer, B., C. Gillott, and M. Erlandson. 2003. Conspecific eggs and bertha armyworm, *Mamestra configurata* (Lepidoptera: Noctuidae), oviposition site selection. Environmental Entomology, 32:529-534.

Vasconcellos-Neto, J., and R. F. Monteiro. 1993. Inspection and evaluation of host plant by the butterfly *Mechanitis lysimnia* (Nymph., Ithomiinae) before laying eggs: A mechanism to reduce intraspecific competition. Oecologia, 95:431-438.

Via, S., R. Gomulkiewicz, G. De Jong, S. M. Scheiner, C. D. Schlichting, and P. H. Van Tienderen. 1995. Adaptive phenotypic plasticity: Consensus and controversy. Trends in Ecology & Evolution, 10:212-217.

West-Eberhard, M. J. 2003. Developmental plasticity and evolution. Oxford Univeristy Press, New York. 794 pp.

Zalucki, M. P., A. R. Clarke, and S. B. Malcolm. 2002. Ecology and behavior of first instar larval Lepidoptera. Annual Review of Entomology, 47:361-393.

7

The Importance of Phenotypic Plasticity in Herbivorous Insect Speciation

Gazi Görür

Department of Biology, Science and Art Faculty, Nigde University, 51100 Nigde, Turkey, ggorur1@nigde.edu.tr, ggorur@excite.com

A major goal of evolutionary biology is to understand divergence and speciation. Until recently, it was believed that speciation was usually initiated when two populations became reproductively isolated; once gene flow between the two populations was reduced, then divergent natural selection could act to change phenotypes by altering gene frequencies to the point where hybridization was no longer possible (Mayr 1963; Rice 1987). However, we now realize that speciation can follow a different pathway, one that starts not with reproductive isolation, but with phenotypic plasticity.

In this chapter, I examine how phenotypic plasticity can foster speciation in herbivorous insects. Insect herbivores are especially appropriate models for understanding speciation because they are incredibly successful and speciose, representing at least one-quarter of Earth's biodiversity (Wilson 1992). Some insect herbivores appear to be undergoing very rapid diversification (Berlocher and Feder 2002). Their great range in feeding habits and lifestyles allows examination of a variety of evolutionary models. The existence of specialist herbivores on ecological islands (specific plant species or plant parts) lends itself to sympatric speciation, when subpopulations "jump" to different sympatric plant hosts, or when generalist species divide into host races. Finally, it is among herbivorous insects that we find some of the most remarkable examples of phenotypic plasticity.

Phenotypic plasticity (the capacity for individuals of the same genotype to express different phenotypes in different environments) had long been considered of lesser evolutionary importance because of its supposed lack of genetic basis, and because phenotypic plasticity was often assumed to buffer the impact of natural selection, and thus act to constrain speciation. As a consequence, there has been relatively less systematic study to analyze environmental effects on the phenotype and the evolutionary consequences of such plasticity. However, it has become increasingly clear that instead of restraining evolutionary change, phenotypic plasticity may actually foster it. Indeed, phenotypic plasticity may be a fundamental component of evolutionary change (Thompson 1991; West-Eberhard 2003) and may be one solution to the problem of adaptation to heterogeneous environments (Via et al. 1995).

How Phenotypic Plasticity Fosters Genetic Divergence and Speciation

Various models propose how phenotypic plasticity might aid speciation (West-Eberhard 2003); many incorporate the following steps:
1. A subpopulation comes to inhabit a different environment.
2. A different phenotype is expressed in the different environment. This population may or may not have previously undergone selection for phenotypic plasticity or genetic variation for the specific environmentally induced plastic trait.
3. Recurrence in the new habitat allows selection on that new phenotype, leading to adaptive genetic evolution via genetic accommodation, resulting in increased fitness.
4. Reduced gene flow between different populations surviving in different habitats allows greater genetic and phenotypic divergence, leading to speciation.

In this model, environmentally induced phenotypic variation (plasticity), can aid speciation when it makes available a different phenotype upon which selection can act, when condition-specific traits are adaptive in the new environment, allowing individuals to survive under a different set of environmental conditions, and when differences in phenotype lead to reduced gene flow. Note that in this model, phenotypic differences are well established *prior* to reproductive isolation. Hence, in this scenario, speciation starts with phenotypic plasticity, not reproductive isolation.

Evidence for Speciation via Phenotypic Plasticity

The evidence for speciation via phenotypic plasticity is largely indirect, but substantial. Included are the ideas that environments vary, organisms often come to reside in new environments, new environments induce phenotypic changes in organisms, these changes are often profound, affecting morphology, physiology, behavior, and ecology, and that environment-induced phenotypic responses are sometimes adaptive, allowing organisms to survive or even thrive under new conditions. Recurrence in a new environment can lead to genetic adaptation and fixation (components of genetic accommodation), reproductive isolation, and speciation.

Virtually all environments vary physically and biologically, and, as a result, all species face environmental heterogeneity. For example, as one moves from the center of a species' range to its edges and beyond, species-relevant environmental conditions generally become more extreme and more unfavorable. For insect herbivores, food plant resources vary spatially and temporally. Plants differ genetically across their range, producing geographic differences in traits that directly affect insect herbivores, such as form, concentration of toxins, harriness, etc. (Bernays and Chapman 1994). Plants also exhibit phenotypic plasticity. A host plant may exist in a dry, wet, sunny, shady, cold, hot, acidic, basic, or high- or low- competition or predation habitat, inducing in that plant a different morphology and physiology, relevant for herbivores. Likewise, plant parts such as buds, emergent leaves, flowers, and fruits that serve as food for phytophagous insects are often ephemeral or temporally variable, and sometimes unavailable. All of these factors serve to create a highly variable environment for insect herbivores.

In virtually every generation, some individuals are exposed to new environmental conditions as they disperse beyond their primary range or move into different sympatric habitats, or as the environment changes. For example, most herbivorous insects possess a dispersal phase, which can place them in a new climatic environment or on a different plant species. In addition, seasonal changes or changes in competitors or natural enemies can greatly alter the habitat. Hence, a phytophagous insect can easily find itself in a very divergent environment.

Research over the last century clearly demonstrates that the environment can induce developmental changes in organisms, producing individuals of the same genotype that differ in morphology, physiology, or behavior (Schlichting and Pigliucci 1998). These changes can greatly influence the ecology, life history, and survival of individuals, and can be harmful. But in some cases, an environmentally induced change can be beneficial in that it allows individuals to survive and reproduce under new or extreme conditions, leading to the permanent establishment (recurrence) of a population in that new environment. For example the smaller adult body mass that is induced in a new, suboptimal environment, may have any number of advantages in that new habitat. Small body mass might allow faster development. It might result in fewer ovarioles or fewer active ovarioles, and, therefore, fewer eggs, requiring less food to complete oogenesis. Small size may allow individuals to complete their development on small resources, such as small seeds. It may allow individuals to better hide, or it may make them unprofitable to large predators, and insufficient for large parasitoids. Small size, and the resulting ability to shelter in small crevices, may reduce the chance of being carried away in windy environments or in flowing streams. It may allow individuals to move freely among spines or plant hairs. Hence, an environment-induced change in phenotype may have beneficial consequences. It may allow a population to inhabit or even prosper in a new environment, without the necessity of genetic change.

Phenotypic plasticity is common in herbivorous insects, and includes environmentally induced differences in morphology, physiology, behavior, life history, and ecology. These plastic responses are very often adaptive. A good example is seen in grasshoppers, where diet differences induce a variety of morphological effects. For example, *Melanoplus femurrubrum* De Geer grasshoppers raised on course, fibrous food develop larger heads and mandibles, and greater mandibular musculature than individuals given softer food (Thompson 1992). In *Melanoplus differentialis* (Thomas), a low-protein diet induces a larger gut, which presumably increases digestion and assimilation, and thus compensates for the poor diet (Yang and Joern 1994). Food type can even influence grasshopper sensory capabilities. Rearing *Locusta migratoria* (Linnaeus) or *Schistocerca americana* (Drury) on artificial diet caused a reduction in antennal sensillium numbers compared with plant-reared individuals (Rogers and Simpson 1997; Bernays and Chapman 1998). Diet can also induce behavioral plasticity in grasshoppers and other phytophagous insects, such as when toxic or nutritionally deficient diets induce host-plant switching (Muralirangan et al. 1977).

A particularly interesting and complex case of phenotypic plasticity is observed in some grasshoppers that respond to environmental cues associated with greater intraspecific competition (Uvarov 1966, 1977; Whitman 1990; Roessingh et al. 1998; Applebaum and Heifetz 1999; Simpson et al, 2001). In locusts, such as *Locusta migratoria* and *Schistocerca gregaria* (Forskal), increased population density increases tactile, pheromonal, and visual stimuli, which induces profound phenotypic changes in morphology, physiology, behavior, and life history. Density related stimuli act to turn the normally solitary grasshoppers into a gregarious form with longer wings, different color, reduced development time, altered fecundity, larger flight muscles, a propensity for aggregation and migration, and greater diet breadth and feeding rate (Uvarov 1966, 1977; Whitman 1990). This plastic response is thought to be adaptive, when it allows aggregating individuals to swamp the predatory response of local predators, and fosters escape from declining habitats via mass migration to regions of rainfall.

Another classic example of highly evolved and adaptive phenotypic plasticity is seen in aphids, which show plasticity in wings, migration, host plant utilization, way of reproduction, and production of soldiers. For example, parthenogenetic females produce either winged or wingless forms in response to host plant quality changes and changes in other environmental factors. During the summer the apterous form of the English grain aphid, *Sitobion avenae* Fabricius, feeds on the flower heads of several species of grasses that provide high quality nutrition for only a short period. Aphids apparently benefit from their wingless, parthenogenetic phase, which gives them some of the highest population growth rates in insects (Dixon 1987). During this period, females have telescoping generations in which parthenogenetic females carry not only the embryos of their daughters but also their granddaughters. Subsequently, the alatae form develops in response to crowding and decrease in host quality (Watt and Dixon 1981). Similar results were shown for the pea aphid *Acyrthosiphon pisum* (Harris) (Shaposhnikov 1965) and the rosy apple aphid *Dysaphis devecta* (Walker) (Forrest 1970).

In aphids, the form of reproduction is plastic, and is determined by changes in host plant quality, day length and temperature (Dixon 1998). Huxley (1858) showed that as long as aphids were kept warm and well fed, they would continue to reproduce parthenogenetically without deterioration. This is supported by the existence of many anholocyclic aphid species. Aphids living in continuous darkness with a relatively constant temperatures and often at considerable depths on the roots of their host plants still produce sexual morphs as the plants become dormant. *Aphis farinosa* Gmelin and *Dysaphis devecta* live above ground, but produce sexual morphs when night length is at its shortest and responding to the cessation of shoot growth of their host plant (Forrest 1970). Whatever the mechanism, plasticity in sexual reproduction enables aphids to synchronize their life cycles with the growth and development of host plants and changes in environmental conditions.

Some aphid species also show a remarkable plasticity in body form and function by producing soldiers that defend the aphid colony from natural enemies. The production of soldiers is sometimes driven by external factors such as day length, temperature, season (Sunose et al. 1991; Tanaka and Ito 1994), host plant (Pike et al. 2004), or crowding and colony size (Sakata et al. 1991; Schültze and Maschwitz 1991; Stern and Foster 1996; Shibao et al. 2003).

Multivoltine insects living in seasonal habitats also often evolve adaptive plasticity that allows individuals to overcome the unfavorable season by switching from an active form to a diapausing or estivating form (Tauber et al. 1986). Switching is often anticipatory. That is, the underlying hormonal changes that ultimately induce the diapausing phenotype are set in motion well before the appearance of the harmful environmental condition, allowing the insect to make the phenotype change before the environment declines. This plasticity requires the evolution of sensory systems that can monitor an environmental cue that precedes and predicts the coming change, and can translate that cue into a physiological/developmental response, such as a change in hormone level. Diapause in multivoltine species is usually facultative and occurs in response to an environmental trigger such as change in photoperiod, temperature, host plant quality, etc.(Hunter and McNeil 2000). Diapause plays an important role in both promoting seasonal synchrony and allowing insects to exist over wide geographical ranges (Bale et al. 2002). There is considerable evidence that environment-induced plastic diapause responses are variable and heritable (Tauber et al. 1986; Nylin 1988; Roff and Bradford 2000).

Insects also show environmentally induced plasticity to pheromones. For example, worker honeybees, *Apis mellifera* Linnaeus, raised in the spring, were more responsive to the queen's mandibular gland pheromone than those reared in the fall, indicating a seasonal phenotypic plasticity (Pankiw et al. 2000).

Many other examples of phenotypic plasticity in herbivore insects exist, including changes in body size, color, wing length, allometry, development, behavior, etc. Hence, it is clear that the environment can greatly influence phenotype, and that insects with identical genomes, can come to have very different phenotypes if subjected to different environmental conditions. Such plasticity is common in insect herbivores and other animals, and has important consequences for speciation.

Selection for Phenotypic Plasticity

As discussed above, most animals, including herbivorous insects, live in spatially variable environments. A phenotype that is optimal in one environment or on one host plant, may well be suboptimal in a different environment or on another host plant. How a population responds evolutionarily to environmental fluctuation may depend on the coarseness of the fluctuating variable (Butlin 1987, 1995; Bush 1975, 1993). Genetic adaptation, such as the development of genetic polymorphism or host or geographic races, would be predicted in course-grain environments where individuals in a given population would only encounter one environment type. In contrast, we would expect phenotypic plasticity to evolve in fine-grain environments where individuals experience multiple environments during their life.

In the same way, migration between heterogeneous environments might play an important role promoting the evolution of phenotypic plasticity (Scheiner 1998; Sultan and Spencer 2002). When environments within the range of a species differ, it may be unlikely that any single genotype will confer high fitness in all situations. In this case phenotypic plasticity provides increased environmental tolerance (Via et al. 1995) and a greater capacity to buffer environmental stress (Thompson 1991).

Two aspects of phenotypic plasticity facilitate phenotypic change (West-Eberhard 1989): a) the occurrence of environment-sensitive expression of phenotypes, and b) the capacity for immediate correlated shifts in related traits. In the first case, the new phenotype always co-occurs with the new environment, which allows alternative phenotypes to be profitably associated with particular conditions; individual traits can be positively selected even when an established phenotype is more efficient in most situations or most individuals (Dawkins 1980; Eberhard 1982). After selection, the new phenotype is expressed in conditions where it is likely to be more advantageous (Shapiro 1978). In the second case, the environment may alter numerous related features simultaneously, presenting a new expression of related traits for evolution to act on. For example, Bernays (1986) demonstrated how differences in diet in the grass-feeding caterpillar *Pseudaletia unipuncta* (Cusson and McNeil) influences both morphology and function. Individuals reared on hard grass developed heads with twice the mass of those fed on soft, artificial diet, and also had greater mandibular strength and ability to feed on course food.

Plastic responses to a fluctuating environment may maintain or even generate additional genetic variability within population. Callahan et al. (1997) proposed that selection for appropriate responses to environmental fluctuations may produce a diversity of genetic and developmental solutions to environmental challenges. Hence, phenotypic plasticity may act as a diversifying factor in evolution, contributing to the origin of novel traits and to altered direction of change (West-Eberhard 1989, 2003). By buffering the population from natural selection, phenotypic plasticity might maintain or even increase genetic variability within a population (Wright 1931, Gillespie and Turelli 1989). In contrast, some authors have suggested that the buffering capacity of phenotypic plasticity reduces the opportunity for natural selection to generate or maintain genetic variation, leading to reduced genetic variation and divergence.

Genetic Accomodation

Phenotypic plasticity produces individuals of the same genotype that can vary in morphology, physiology, behavior, ecology, and life history. Such variability may aid dispersal into new environments when the plastic response induced in a new environment allows or increases survival and reproduction in that environment. Recurrence is an important step in speciation via phenotypic plasticity, because it allows natural selection to begin to alter gene frequencies through genetic accommodation. Genetic accommodation can be defined as the gene frequency change due to selection on variation in the regulation, form, or side effects of the novel trait in the subpopulation of individuals that express the trait (West-Eberhard 2003). In addition, West-Eberhard predicts that environmentally triggered novelties may have greater evolutionary potential than mutationally induced ones, for at least two reasons: First, an environmental factor tends to influence large groups, whereas a mutation initially affects only one individual. Such intraspecific recurrence enables the trait to be tested in many different genetic backgrounds. Second, as discussed above, an environmentally triggered novelty is automatically associated with a particular environmental situations. Thus, such novel phenotypes are more subject to consistent selection and directional modification than are mutationally induced novelties.

Genetic accommodation also improves a novel phenotype in at least three ways: a) by adjusting regulation, to change the frequency of expression of the trait, b) by adjusting the form of the trait, improving its efficiency and integration, c) by reducing disadvantageous effects. Genetic accommodation of traits was implicit in Baldwin's (1902) hypothesis, which was later known as the *Baldwin effect*. The Baldwin effect is the idea that phenotypic accommodations to variable or extreme situations may affect the direction of genetic evolution under natural selection, allowing survival in an unusual environment, and permitting time to favor adaptation to it. The Baldwin effect generally has been synonimized with genetic assimilation. Genetic assimilation is a process by which a novel phenotypic character, which is initially produced only in response to environmental cues, becomes, through a selection, taken over by genes and therefore this novel phenotype is found even in the absence of the environmental influence that was necessary at the initial expression. West-Eberhard (2003) pointed out that although genetic accommodation can include some aspects of both Baldwin effect and Genetic assimilation, there are some basic differences, which are; a) Genetic accommodation refers to adjustments of form as well as frequency of expression and selection may simultaneously imrove the shape of a novel phenotype and change its frequency of occurrence. Genetic assimilation and the Baldwin effect refer only to change in regulation. b) Genetic accommodation includes novel traits originated by either genetic mutation or environmental induction. Genetic assimilation and Baldwin effect apply only to phenotypes that are originally environmentally induced, and then become increasingly genetically influenced, c) Genetic accommodation may increase phenotypic susceptibility to environmental influence whereas genetic assimilation may decrease environmental specificity of expression. d) Genetic accommodation can modify initially deleterious phenotypes whereas genetic assimilation and Baldwin effect refer only to evolutionary change in positively selected traits. Hence, genetic accommodation may play an important role in the persistence and evolutionary consequence of phenotypically plastic traits.

Reproductive Isolation, Host Plant Races, and Speciation

Phenotypic plasticity, recurrence, and genetic accommodation can lead to reduced gene flow between two populations living under divergent environmental conditions. Reduced gene flow facilitates genetic divergence. The initial viability of a novel trait is increased by adaptive plasticity, followed by genetic accomodation due to selection which may affect the regulation and form of the novel trait (West-Eberhard 2003). If expression of any life history trait is associated with particular environmental conditions, recurrence might cause differences in assortative mating and create partial reproductive isolation. When the frequency of a novel trait increases in a population in a novel environment, gene flow may be reduced and genetic divergence may be increased between two populations.

Insect herbivores may be especially susceptible to reproductive isolation and speciation via a unique type of phenotypic plasticity known as the "Hopkin's host selection principle." In this process, insects that are exposed to a particular host come to prefer that or related hosts. This process could act to keep a population on a new host. For example, Copp and Davenport (1978) reported a three-fold increase in the relative number of *Agraulis vanillae* (Linnaeus) eggs laid on *Passiflora mollissima* (Kunth) vs. *P. caerulea* (Linnaeus), when the former was the larval host, although these findings were challenged by some researchers. Switches in host preference are important, because recurrence on the new host could allow rapid adaptive genetic evolution. For example, Shaposhnikov (1965, 1966) showed that after only eight parthenogenetic generations the aphid *Dysaphis anthrisci* Börner was able to survive and reproduce on *Chaerophyllum maculatum* Linnaeus, a plant that previously had caused nearly 100% mortality. Likewise, green peach aphids *Myzus persicae* (Sulzer) had higher survival and slightly higher fecundity after being reared for several generations on an artificial diet with coumarin, a common plant secondary chemical, which normally caused high mortality (Görür and Mackenzie 1998). After this selection, these aphids could survive at even higher doses of coumarin.

Phenotypic plasticity allows for colonization and succcess in a novel habitat and other forces, such as induced host choice and genetic accommodation, cause restricted gene flow between the old and new habitats. If plastic organisms are restricted to or favor the novel habitat, a new host race may be formed (Guldemond 1990b; Agrawal 2001). Host-plant shift and the development of host races are especially important to the concept of speciation via phenotypic plasticity. Indeed, host shifting is extremely common in insect herbivores, and may be the primary means of speciation in this group (Diehl and Bush 1984, Strong et al. 1984, Futuyma and Peterson 1985). Treehoopers in the *Enchenopa binotata* (Say) species complex provide a model for speciation in sympatry through shifts to novel host plants; host fidelity in mating and oviposition on new hosts reduces gene flow among host-shifted populations (Wood et al. 1999; Sattman and Coccroft 2003). Plasticity in male mate behaviour on host and non-host population may have the potential to reduce interbreeding between populations that use different host plant species. Once a population becomes established in a new environment, rapid genetic adaptation can occur. Vanbergen et al. (2003) demonstrated how recent host shift resulted in considerable differences in various life history traits such as larval survival and adult size in *Operophtera brumata* (Linnaeus).

In aphids, several studies suggest that variation between species or populations in response to selection by diversity of plant surfaces has resulted in morphological specialization for grasping and locomotion (Moran 1987, Kennedy 1986) and effective probing and penetrating of plant surfaces for food or oviposition (Dixon 1987, Carroll and Boyd 1992).

Morphological characters of aphids have important roles in the adaptation to different host plants. Moran (1986) showed that *Uroleucon* species have longer hind tarsi when using hairy plant species and shorter hind tarsi when using smooth plant species. She suggests that short tarsi help to prevent entanglement in trichomes whereas long tarsi provide a better grasp. In contrast, Eastop (1985) noted that aphids on hosts with sticky exudates had short tarsi, and in extreme, a species described on *Grindelia squarrosa* Pursh has no tarsi (Gillette 1911). Bernays (1991) suggested that walking on "tip toe" may reduce work and allows fast movement. Also, aphids specialized on *Tilia* spp. with dense stellate hairs had an elongate rostrum (Carter 1982). Host-specific morphology has been described for a specialist oak aphid, *Myzocallis schreiberi* Hille Ris Lambers (Kennedy 1986, Southwood 1986), which possesses a terminal tarsal pair of claws and a pair of flexible spatulate empodia, which allow a good grasp on the hairs of the densely pubescent leaves of *Quercus ilex* Linnaeus.

Dixon et al. (1995) pointed out that aphid size is strongly correlated with the length of its proboscis which is determined by the depth to which aphid has to probe plant tissues in order to feed. The depth of phloem varies with plant species and the position of the aphids on the host. Thus the depth of the phloem element results in a particular optimum aphid size. Carroll and Boyd (1992) also demonstrated that the length of the beak of *Jadera haematoloma* Herrich and Schaeffer differed related with the host plant differences over the last 20-50 years in Florida.

The mean value of any morphological character expressed on two host plants might be expected to reach an optimum, as a result of stabilizing selection (Via and Lande 1985). This might result in host–dependent allometric growth, resulting in shape changes that adjust morphology to the contemporary host plant (Blackman and Spence 1994). Weisser and Stadler (1994) proposed that the optimal decision on which morph to produce depends not only on the current quality of the plant where the aphid feeds but also on the predicted future quality of this plant, in terms of phenotypic plasticity.

Phenotypic plasticity not only allows the colonisation of a new host plant, but it is also a mechanism that may reinforce differences in the ecological context of the ancestral and new host. When new host preference arises in a population via learning or conditioning, and mating is linked to host choice, this can decrease gene flow, speeding the establishment of genetically distinct host races (Maynard Smith, 1966, Jaenike 1981, Rice 1987). Several researchers (Bush 1975, Müller 1985, Diehl and Bush 1989, Tauber and Tauber 1989) have pursued this outlook. They have argued that host race formation represents the initial stages of sympatric speciation, in which the incipient species are isolated not by geographic barriers but by the choice of host plants on which mating occurs. If a herbivore species exhibits a plasticity in choice and displays induced performance on a novel host plant, colonization of this host plant may give rise to offspring which prefer that host plant over others and in the absence of major genetic changes phenotypic host races have occurred (Mackenzie and Guldemond 1994).

Individuals on the new host plant are faced with a different set of interactions (Leclaire and Brandl 1994), which may result in new genomic organization (Holloway et. al.1990). A population that inhabits heterogeneous environments may be selected to evolve a genetic constitution that allows phenotypic variability to adjust to different environments so as to increase its fitness (Zhivotovskoy et al. 1996). The formation of new host races especially on chemically and morphologically different plants or on cultivated plants may initially require small changes in the genome. Selection may eventually lead to reorganization of at least a portion of the genome in the new host races. It has been suggested that during the formation of a new host race about 10% of the insect genome is altered (Bush 1975). When gene frequency changes in population over a period of time following the initial colonization that could allow the insect to exploit the new host in a better way. Such changes may increase the degree of reproductive isolation between the old and new population over time and phenotypic plasticity may reinforce differences in the ecological context of ancestral and new host (Leclaire and Brandl 1994) and might lead to host race formation. Host race formation is a valuable model for studying phenotypic plasticity and sympatric speciation. Sympatric speciation in phytophagous insects is facilitated by the ecological and genetic potential to shift hosts and form host races (Bush 1975, Abrahamson and Weis 1997). Given sufficient variation for phenotypic plasticity, host race formation can be favored and maintained in phytophagous insects (Peppe and Lomonaco 2003). Walsh (1864) proposed that in phytophagous insects a host shift caused by changing host preference may give rise to new host races. He reasoned that there is sufficient "phytophagic isolation" (host induced specificity) to allow the adaptive evolution of new species in the absence of geopraphical isolation. It has been shown that host-specific selection pressures may result in partial or complete genetic isolation of host associated subpopulations of herbivorous insects resulting in sympatric host race formation (Craig et al. 1993, Feder et.al. 1996, Mopper 1996). Despite that, the evolutionary mechanisms of sympatric speciation through host race formation in phytophagous insects have long been suspected.

The most widely acclaimed example of sympatric host race formation is that in *Rhagoletis pomonella* (Walsh), which has apple and hawthorn races. Host-associated traits have been shown to be primarily responsible for isolating *Rhagoletis pomonella* groups. *Rhagoletis* adults court and mate exclusively on or near the fruit of their host plants (Prokopy et al. 1971, 1972); differences in host preference behaviors translate directly into mate choice decisions and pre-mating isolation (Prokopy et al. 1988, Feder et al. 1994). Adult individuals reared from hawthorn have slightly longer ovipositors than those associated with apple (Bush 1969) but there was still gene flow between two races. Apparently, gene flow between apple and hawthorn races has been reduced gradually, about 4-6 % per generation (Feder et al. 1994, Feder 1996). Berlocher (1999) compared 17 allozyme loci in 21 populations between the flowering dogwood fly and its closest relative the apple maggot fly, *R. pomonella* and showed significant frequency differences. Later, relatively low gene flow was demonstrated between the dogwood fly and *R. pomonella* populations. The races differ in host selection behaviour, life history timing (especially diapause timing) and allozyme frequencies (McPheron et al. 1988, Prokopy et al. 1988, Feder et. al. 1993, Feder 1996). Evidence suggests that the apple maggot, *Rhagoletis pomonella* is undergoing sympatric speciation in the process of shifting and adapting to a new host plant (Feder et al. 2003) via phenotypic plasticity. In contrast to *R. pomonella* groups, *R. cerasi* Linnaeus can not form sympatric host races as there is a lack of genetic variation between two populations on different host plants, although they expressed host-associated differences for one loci (Schwarz et. al.2003).

Many aphid species are composed of a number of host races, adapted to and preferring different host plants, and these races may occur sympatrically (Eastop 1973, Müller 1985, Guldemond 1990a, Mackenzie 1996) via phenotypic plasticity. Such races may also arise suddenly on cultivated fields and rapidly cause crop damage (Eastop 1973). It has been shown that once aphid species shift host plants, they quickly establish on the new host and rapidly increase the frequency of novel traits. Despite gene flow between two populations on different host plants, genetic differences between host races in aphids appear to quickly evolve (Müller 1982, Jorg and Lampel 1996). Masutti and Chavighny (1998) showed that host plants have acted as a strong selective forces leading to the genetic differentiation of *Aphis gossypii* Glover on different host types as a host races. De Barro et al. (1995) also showed similar differences for the grain aphid *Sitobion avenae* related with host plants type. Shufran et al. (2000) demonstrated the divergence in mitochondrial DNA sequence among greenbug biotypes and these biotypes may have diverged as host adapted races on wild grasses. *Schizaphis graminum* Rondani biotypes are a mixture of genotypes belonging to three clades and may have diverged as host adapted races on wild grasses. Godfrey (2001) showed a significant amount of phenotypic plasticity of insecticide susceptibility in genetically identical cotton aphids. This plasticity has important implications for both cotton aphid physiology and management.

Although there are severall doubts about whether or not the initial divergence between *Acyrthosiphon pisum* races was sympatric or allopatric (Via 2001), the pea aphid, *A. pisum* consists of a number genetically distinct host races or biotypes, which have specialized on certain species of papilionaceous plants (Müller 1982, Via 1991, 1999, Sandström 1996, Hawthorne and Via 2001). Müller (1985) defined them as sympatric races isolated from each other mainly by strict host preference and by hybrid inferiority and Via (1999) demonstrated differences in allozyme allele frequencies. Komazaki (1986) showed genetic differentiation and the probability of hybridization between *Aphis citricola* van der Goot populations on *Citrus unshiu* Marcovitch and *Spiraea thunbergii* Sieb, Dres and Mallet (2002) mentioned them as possible host races. The gall-forming aphid, *Tetraneura yezoensis* Matsumura formed host races galling on *Ulmus davidiana* Planchon and *U. laciniata* Trautvetter and these indicated that aphid populations on two plants are genetically differentiated in physiological traits (Akimoto 1990). Guldemond (1990a, b) showed that two host races exhibited in *Cryptomyzus galeopsidis* Kaltenbach are restricted to *Ribes rubrum* Linnaeus and *R. nigrum* Linnaeus and then they differed genetically at allozymes.

Host races are also common in other insect groups. For example, races of the obscure scale, *Melanaspis obscura* Comstock, infest various species of oak, and have differences in the timing of male development, which is synchronized with host phenology, preventing interbreeding (Miller and Kosztarab 1979). Claridge et al. (1977) pointed out that populations of the univoltine leafhopper *Oncopsis flavicollis* Linnaeus may be subdivided into host races by differences in an oviposition differences There were host races of *Eurosta solidaginis* Fitch feeding on *Solidago altissima* (Linnaeus) and *S. gigantea* Aiton with strong ecological, behavioral and genetic differences driven by phenotypic plasticity (Waring et al. 1990, Brown et al. 1996). There were also host races of the goldenrod elliptical gall moth *Gnorimoschema gallaesolidaginis* (Riley) on Solidago plants. Nason et al. (2002) showed consistent and significant differences between the goldenrod elliptical gall moth host races. Rapid host race formation has occurred in a milkweed beetle, *Tetraopes tetrophthalmus* Forster, where a reduced probability of mating between populations on *Asclepias verticullata* Winterberry and *A. syriaca* Linnaeus was found (Price and Wilson 1976). There are also host races in the Lepidopteran *Papilio demodocus* Esper, which infects citrus and Umbelliferae species. Two distinct larval morphs, whose expressions would be controlled by a single gene, occur on different host and the forms interbreed in the laboratory and there are no differences in survival when reared on host plants reciprocally (Clarke et al. 1963). Nice and Shapiro (2001) showed restricted gene flow between butterfly *Mitoura nelsoni* (Boisduval) and *Mitoura muiri* (Edwards) populations and these restrictions may result from the adults' habit of mating preferentially on different host trees and through phenological differences in the timing of reproduction. They concluded that divergence among these populations has occurred very recently and sympatric divergence of host races is a plausible explanation. The brown planthopper *Nilaparvata lugens* (Stal) using *Leersia hexandra* Swartz and *Oryza sativa* Linnaeus expressed heritable differentiation in a number of life history traits (Sezer and Butlin 1998 a, b) and they might be accepted as host races or species (Claridge et al.1988, Berlocher and Feder 2002). It has been demonsrated that *Viburnum* feeding species of the treehopper *Enchenopa binotata* (Say) showed potential for better feeding on alternative *Viburnum* species and these may lead to formation of host races and these races can be identified by the allozyme differences and nymphal morphological characters (Pratt and Wood 1992, Tilmonn et al. 1998, Wood et al. 1999).

Groman and Pellmyr (2000) demonstrated morphological, phenological and significant genetic differences in a yucca moth population, *Prodoxus quinquepunctellus* Chambers, when colonized on a novel host. They postulated that interspecific differences at the macroevolutionary level in *Prodoxidae* can accumulate rapidly as a result of novel host plant colonization. The host races of larch budmoth *Zeiraphera diniana* Guenee differ in larval colour morphs and these differentiations are also associated with host plants features (Baltensweiller et al. 1977). Emelianov et al. (2001, 2003) showed that phenotypic plasticity in host choice behavior promotes reproductive isolation between host races of the larch budmoth *Zeiraphera diniana*, feed on *Larix decidua* Miller and *Pinus cembra* Linnaeus and *Z. diniana* races differ at three identified allozyme loci. The small ermine moth, *Yponomeuta padellus* Linnaeus, has a number of host races infesting the species of *Rosaceae* and *Salix* (Menken 1981). There are also host races of the small ermine moth feeding on different host plants. Recent studies carried out by Raijmann and Menken (2000) demonstrated the differences in the allozyme frequency in the small ermine moth, *Yponomeuta padellus,* collected on *Crataegus* spp. and *Prunus* spp. from several sites. Bourguet et al. (2000) found that populations of the *Ostrinia nubilialis* Huebner on cultivated maize *Zea mays* Linnaeus are genetically different from populations colonize on the wild host sagebrush, *Artemisia* sp. Diegisser (2004) showed host associated genetic and morphological differences in *Tephritis bardanae* Schrank and it can be considered as sympatric host races due to host associated phenotypic plasticity in oviposition behavior and morphological traits.

Kreslavsky et al. (1981) demonstrated the lower probability of the hybridization between birch and willow races of the chrysomelid beetle *Lochmaea caprea* Linnaeus. Host race formation in Chrysomelid beetle was driven by phenotypic plasticity in host choice behavior. Kaneko (2002) showed that developmental plasticity induced by interaction leads to phenotypic differentiation leading to hybrid sterility and eventually reproductive isolation. Heitland and Pschorn-Walcher (1992) showed that *Platycampus luridiventris* Fallen host races on *Alnus glutinosa* Linnaeus and *A.incana* Linnaeus differ in larval morphology and female oviposition preference and these differences are probably genetic. Craig et al. (1997) demonstrated that each host race has a strong preference for oviposition on its own host plant where survival is significantly higher than on the alternate host plant. The gall-forming midge, *Asphondylia borrichiae* Rossi and Strong may also form host races on three different coastal plants and these race formations may be determined by parasitism and plant phenology (Rossi et al. 1999).

Herbivore speciation via phenotypic plasticity may even occur within a single plant host. Despres and Jaeger (1999) showed that *Chiastocheta* spp. feeding on the same host, *Trollius europeaus* Linnaeus, had speciated sympatrically into six species. They postulated that in contrast to speciation via host shifts, *Chiastosheta* speciation took place within the host plant. Reproductive isolation between species on the same host was driven by differences in oviposition behavior and also sexual selection. There are studies on fig pollinators and their parasites examine how modes of speciation may be affected by interspecific interactions. Weiblen and Bush (2002) pointed out that sister species of fig parasites attack the same host, but show different ovipositor lengths, which indicates that parasites speciation could result from divergence in the timing of oviposition related with fig development. Their results highlighted the importance of interspecific interactions in sympatric speciation in plant-feeding insects and their parasites.

Strong intraspecific competition may also favour host race formation and sympatric speciation through host shifts in herbivorous insects. Craig et al. (2000) showed that the preference for rapidly growing ramets and intraspecific competition may have facilitated a host shift in *Eurosta solidaginis*.

It is also possible that insect species whose life histories are closely related with other insects may show differences in response to a shift in resource use and specialized natural enemies may diversify in parallel with their herbivorous prey after the herbivores complete host shift. The tumbling flower beetle *Mordellistena convicta* LeConte, which is a predator of *Eurosta solidaginis*, has formed host races in response to a host plant shift and host race formation by its prey (Abrahamson et al. 2001, Eubanks et al. (2003). Cronin and Abrahamson (2001) reached the opposite conclusion for the parasitoid wasp, *Eurytoma gigantea*, which parasited two host races, but did not express any ecological, behavioral or genetic differences.

It is clear that host race formation via phenotypic plasticity is common in herbivorous insects and host races in plant feeding insects can be considered as evidence for the probability of sympatric speciation. Dres and Mallet (2002) pointed out that sympatric speciation via host race formation was likely in nature because: a) assortative mating via pleiotrophy was likely to occur during the evolution of host choice and host adaptation, b) it is theoretically plausible and empirically demonstrated that host races can maintain multiple genetic differences in linkage disequilibrium and c) divergence is probable in cases where linkage disequilibrium existed between divergent traits.

Sympatric speciation through host race formation on novel host plant has been suggested as a major factor in the extensive diversification of phytophagous insects (Bush 1975). Numerous studies support this idea (Diehl and Bush 1984, Tauber and Tauber 1989, Carrol and Boyd 1992, Sembene and Delobel 1998). Therefore it can be considered that phenotypic plasticity is potentially important in giving rise to "phenotypic host races" and hence facilitating sympatric speciation.

Conclusion

Herbivorous insects have long been thought to be good candidates for sympatric speciation because of their intimate and frequently highly specialized relationship with their host plants (Via 2001) and recently it has been considered that phenotypic plasticity has a crucial importance in sympatric speciation. West-Eberhard (2003) pointed out that phenotypic plasticity can facilitate evolution by immediate accommodation and exaggeration of change. In order to clarify its importance in the speciation process, phenotypic plasticity studies in herbivorous insects should be carefully linked to their ecological context and differences between their host plants to facilitate host race formation. It can be shown that many of these races may be spreading toward speciation as they continue to adapt to their host plants (Berlocher and Feder 2002). Although allopatry may be the prime method for speciation in herbivorous insects, sympatric speciation through host race formation is also common.

Continuing research on phenotypic plasticity in herbivorous insects will further illuminate its role in speciation. Phenotypic plasticity in phytophagous insects in terms of having ability to express alternative performance, morphology and behaviour is considered to be crucial to the comprehension of evolutionary mechanisms, especially host race formation and then sympatric speciation (Bush 1969, 1975, De Jong 1995, Wool and Hales 1997). Herbivorous insect genotypes show substantial genetic variation within species for this plasticity. The differential response of one genotype to different hosts and the variation in this response between genotypes provide strong evidence that phenotypic plasticity may facilitate host race formation. The fact that there is often at least partial gene flow between host races, suggests that the main reason for host race formation is not reproductive isolation, but phenotypic plasticity. When the frequency of novel traits increases and becomes permanent on the novel host plant, it increases reproductive isolation between host races. The result might be genetic accommodation and divergent change. Therefore, host races in plant feeding insects have been used as evidence for the probability of sympatric speciation (Görür 2000, Dres and Mallet 2002) and sympatric speciation in herbivorous insects has become increasingly accepted in the past decade (Via 2001). Recent field experiments indicate that selection pressures imposed by host-plants can promote rapid adaptive evolution in natural insect populations and evolution of plant related adaptations that may eventually lead to reproductive isolation between races (Mopper 1996). In fact, host races representing incipient stage of sympatric speciation and host associated species is a final product (Berlocher and Feder 2002). Host plant adaptation potentially leading to host race formation can proceed extremely rapidly and there are clear cases that have produced host adapted races rapidly within the last century. Although the number of host races via phenotypic plasticity in phytophagous insects discovered so far is small, the number of insect systems that may conceal them is very large, and the process may continue (Berlocher and Feder 2002; Dres and Mallet 2002).

New host race formation via phenotypic plasticity may influence disease transfer for agricultural crops. Therefore, study of phenotypic plasticity in host race formation and sympatric speciation in herbivorous insects must include the response of associated organisms (Eubanks et. al. 2003) and their implications for agriculture and conservation.

Finally, present evidence and growing interest suggest that phenotypic plasticity is crucial to the evolutionary processes of herbivorous insects. The theoretical and empirical studies given in this chapter indicate how phenotypic plasticity facilitates speciation phenotypic plasticity, may be established prior to reproductive isolation. Hence, speciation can start with phenotypic plasticity, not with reproductive isolation.

Acknowledgment

I gratefully thank Prof. Dr. Doug Whitman and Assist. Prof. Dr. Metin Sezer for their kind suggestions on this chapter.

References

Abrahamson, W. G. and A. E. Weis. 1997. Evolutionary ecology across three trophic levels: Goldenrods, Gallmakers and Natural enemies. pp. 455, Princeton, NJ: Princeton University Press.

Abrahamson, W. G. and M. D. Eubanks, C. P. Blair, and A. V. Whipple. 2001. Gall flies, Inquilines and Goldenrods: A model for host race formation and sympatric speciation. Amer. Zool. 41: 928-938.

Agrawal, A. A. 2001. Phenotypic plasticity in the interactions and evolution of species. Science, 294: 321326.

Akimoto, S. 1990. Local adaptation and host race formation of a gall-forming aphid in relation to environmental heterogeneity. Oecologia, 83: 162-170.

Applebaum, S.W. and Y. Heifetz. 1999. Density-dependent physiological phase in insects. Annual Review of Entomology, 44: 317-341.

Baldwin, J. M. 1902. Development and Evolution. New York.

Bale, J.S. and G. J. Masters, I. D. Hodkinson, C. Awmack, T. M. Bezemer, V. K. Brown, J. Butterfield, A. Buse, J. C. Coulson, J. Farrar, J. E. G. Good, R. Harrington, S. Hartley, T. H. Jones, R. L. Lindroth, M. C. Press, I. Symrnioudis, A. D. Watt, and J. B. Whittaker. 2002. Herbivory in global climate change research: direct effects of rising temperature on insect herbivores. Global Change Biology, 8: 1-16.

Baltensweiler, W. and G. Benz, P. Bovey, and P. Delucci. 1977. Dynamics of larch budmoth populations. Annual Review of Entomology, 22; 79-100.

Berlocher, S. H. 1999. Host race or species? Allozyme characterization of the "flowering dogwood fly" a member of the *Rhagoletis pomonella* complex. Heredity, 83: 652-662.

Berlocher, S. H. and J. L. Feder. 2002. Sympatric speciation in pytophagous insects: Moving beyond controversy? Annual Review of Entomology, 47: 773-815.

Bernays, E. A. 1986. Diet-induced head allometry among foliage-chewing insects and its importance for graminivores. Science, 231: 495-497.

Bernays, E. A. 1991. Evolution of insect morphology in relation to plants. Philosophical Transactions of the Royal Society of London. Series B, 333: 257-264.

Bernays, E. A. and R. F. Chapman. 1994. Host plant selection by phytophagous insects. Chapman & Hall, New York-London, 314 pp.

Bernays, E. A. and R. F. Chapman. 1998. Phenotypic plasticity in numbers of antennal chemoreceptors in a grasshopper: effects of food. Journal of Comparative Physiology A, 183: 69-76.

Blackman, R. L. and S. M. Spence, 1994. The effects of temperature on aphid morphology using a multivariate approach. European Journal of Entomology, 91: 7-22.

Bourguet, D. and M. T. Bethenod, C. Trouve, and F. Viard. 2000. Host plant diversity of the European corn borer Ostrinia nubilalis: what value for sustainable transgenic insecticidal Bt maize? Philosophical Transactions of the Royal Society of London. Series B, 267: 1177-1184.

Brown, J. M. and W. G. Abrahamson, and P. A. Way. 1996. Mitochondrial DNA phylogeography of host races of the goldenrod ball gallmaker Eurosta solidaginis (Diptera: Tephritidae). Evolution, 50: 777-786.

Bush, G. L. 1969. Sympatric host race formation and speciation in frugivorous flies of the genus Rhagoletis (Diptera: Tephritidae). Evolution 23: 237-251.

Bush, G. L. 1975. Sympatric speciation in phytophagous parasitic insects. pp:187-206. In Price P.W.[ed]. Evolutionary strategies of parasitic insects and mites. Plenum, New York,

Bush, G. L. 1993. Host race formation and sympatric speciation in Rhagoletis fruit flies (Diptera: Tephritidae). Psyche, 99: 335-357.

Butlin, R. K. 1987. Species, speciation and reinforcement. American Naturalist 130: 461-464.

Butlin, R. K. 1995. Reinforcement: an idea evolving. Trends in Ecology and Evolution, 10: 432-434.

Callahan, H. S. and M. Pigliucci, and C. D. Schlichting. 1997. Developmental phenotypic plasticity: where ecology and evolution meet molecular biology. Bioessays, 19: 519-525.

Carroll, S. P. and C. Boyd, 1992. Host race radiation in the soap-berry bug: Natural history with the history. Evolution, 46: 1052-1069.

Carter, C. I., 1982. Susceptibility of Tilia species to the aphid Eucallipterus tilae, 423-433. In Visser, J. H. and A. K. Minks, [eds.]., Proceedings of 5th International Symposium of Insect-Plant Relationships, Wageningen, Pudoc.

Claridge, M. F. and W. J. Reynolds, and M. R. Wilson. 1977. Oviposition behaviour and food plant discrimination in leafhoppers of the genus Oncopsis. Ecological Entomology, 2: 19-25.

Claridge, M. F. and J. Den Hollander, and J. C. Morgan. 1988. Variation in host plant relations and courtship signals of weed associated populations of the brown planthopper, Nilaparvata lugens from Australia and Asia: a test of the recognition species concept. Zoological Journal of Linnaen Society, 35: 79-93.

Clarke, C. A. and C. G. C. Dickson, and P. M. Sheppard. 1963. Larval color pattern in Papilio demodocus. Evolution, 17: 130-137.

Copp, N. H. and D. Davenport. 1978. Agraulis and Passiflora I Control of specificity. Biological Bulletin, 155: 112.

Craig, T. P. and J. K. Itami, W. G. Abrahamson, and J. D. Horner. 1993. Behavioral evidence for host race formation in *Eurosta solidaginis*. Evolution, 47: 1696-1710.

Craig, T. P. and J. D. Horner, and J. K. Itami. 1997. Hybridization studies on the host races of *Eurosta solidaginis*: implications for host shift and speciation. Evolution, 55: 773-782.

Craig, T. P. and J. K. Itami, C. Shantz, W. G. Abrahamson, J. D. Horner, and C. V. Craig. 2000. The influence of host plant variation and intraspecific competition on oviposition preference and offspring performance in the host races of *Eurosta solidaginis*. Ecological Entomology, 25: 7-18.

Cronin, J. T. and W. G. Abrahamson. 2001. Do parasitoids divesify in response to host plant shifts by herbivorous insects? Ecological Entomology, 26: 347-355.

Dawkins, R. 1980. Good strategy or evolutionary stable strategy? In: Sociobiology: Beyond Nature/Nurture. B. W. Barlow and J. Silverberg (eds.), Westview Press, Boulder, pp.331-367.

De Barro, P. J.and T. N. Sherratt, O. David, and N. Maclean. 1995. An investigation of the differential performance of clones of the aphid, *Sitobion avenae*, on two host species. Oecologia, 104: 379-385.

De Jong, G. 1995. Phenotypic plasticity as a product of selection in a variable environment. American Naturalist, 145: 493-512.

Despres, L. and N. Jaeger. 1999. Evolution of oviposition strategies and speciation in the globe-flower flies *Chiastocheta* spp. (Anthomyiidae). Journal of Evolutionary Biology, 12: 822-831.

Diegisser, T. and J. Johannesen, and C. Lehr, and A. Seitz. 2004. Genetic and morphological differentiation in *Tephritis bardanae* (Diptera: Tephritidae): evidence for host-race formation. Journal of Evolutionary Biology, 17: 83-93.

Diehl, S. R. and G. L. Bush. 1984. An evolutionary and applied perspective of insect biotypes. Annual Review of Entomology, 29: 471-504.

Diehl, S. R. and G. L. Bush. 1989. The role of habitat preference in adaptation and speciation. 345-365. *In* Otte, D. and J. Endler. [eds.]. Speciation and its consequences. Sinauer, Sunderland, Massachusetts.

Dixon, A. F. G. 1987. Aphid reproductive tactics, 3-18. *In* Holman, J.and J. Pelikan, A. F. G. Dixon, and L. Weismann [eds.]. Population structure, genetics and taxonomy of aphids and tysanoptera SPB Academic Publishing, The Hague, Netherlands.

Dixon, A. F. G. and P. Kindlmann, and V. Jarosik. 1995. Body size distribution in aphids: relative surface area of specific plant structure. Ecological Entomology, 20: 111-117.

Dixon, A. F. G. 1998. Aphid Ecology. Second edition. Chapman & Hall, N.Y.

Dres, M. and J. Mallet. 2002. Host races in plant-feeding insects and their importance in sympatric speciation. Philosophical Transactions of the Royal Society of London. Series B, 357:471-492.

Eastop, V. F. 1973. Biotypes of aphids. Bulletin of Entomological Society of New Zealand, 2: 40-51.

Eastop, V. F. 1985. The acquisition and processing of taxonomic data, 245-270. *In* Szelegiewicz, H.[ed.]. Proceedings of International Aphidology Symposium of Jablonna 1981, Wroclaw, Poland, Ossolineum.

Eberhard, W. G. 1982. Beetle horn dimorphism: making the best of a bad lot. American Naturalist, 119: 420-426.

Emelianov, I. and M. Dres, W. Baltensweiler and J. Mallet. 2001. Host-induced assortative mating in host races of the larch budmoth. Evolution, 55: 2002-2010.

Emelianov, I. and F. Simpson, P. Narang, and J. Mallet. 2003. Host choice promotes reproductive isolation between host races of the larch budmoth *Zeiraphera diniana*. Journal of Evolutionary Biology, 16: 208-218.

Eubanks, M. D. and C. P. Blair, and W. G. Abrahamson. 2003. One host shift leads to another? Evidence of host arec formation in a predaceous gall-boring beetle. Evolution, 57(1): 168-172.

Feder, J. L. 1998. The apple maggot fly *Rhagoletis pomonella*. Flies in the face of conventional wisdom about speciation, 291-308. *In* Howard, D. J. and S. H. Berlocher. [eds.]. Endless forms: Species and speciation. Oxford University Press, New York.

Feder, J. L. and T. A. Hunt, and G. L. Bush. 1993. The effect of climate, host phenology and host fidelity on the genetics of apple and hawthorn infesting races of *Rhagoletis pomonella*. Entomologia Experimentalis et Applicata, 69: 117-135.

Feder, J. L. and S. B. Opp, B. Wlazlo, K. Reynolds, W. Go, and S. Spisak. 1994. Host fidelity is an effective premating barrier between sympatric races of the apple maggot fly. Proceedings of the National Academy of Sciences of the United States of America, 91: 7990-7994.

Feder, J. L. and S. H. Berlocher, and S. B. Opp. 1996. Sympatric host race formation and speciation in Rhagoletis (Diptera: Tephritidae): a tale of two species for Charles D., 408-441. *In* Mopper, S. and S. Strauss. [eds.]. Genetic Structure in Natural Insect Populations: Effects of Ecology, Life History and Behavior. Chapman & Hall, New York.

Feder, J. L. and J. B. Roethele, K. Filchak, J. Niedbalski, and J. Romero-Severson.2003. Evidence for Inversion Polymorphism Related to Sympatric Host Race Formation in the Apple Maggot Fly, *Rhagoletis pomonella*. Genetics, 163: 939-953.

Forrest, J. M. S. 1970. The effects of maternal and larval experience on morph determination in *Dysaphis devecta*. Journal of Insect Physiology, 16: 2281-2282.

Futuyma, D. J. and C. S. Peterson. 1985. Genetic variation in the use resources by insects. Annual Review of Entomology, 30: 217-238.

Gillespie, J. H. and M. Turelli. 1989. Genotype-environment interaction and maintainance of polygenic variation. Genetics, 121: 129-138.

Gillette, C. P. 1911. A new genus and four species of Aphididae. Entomological News, 22: 440-444.

Godfrey, L. D. and F. J. Fuson. 2001. Environmental and host plant effects on insecticide susceptibility of the cotton aphid (Homoptera: Aphididae). The Journal of Cotton Science, 5: 22-29.

Görür, G. and A. Mackenzie. 1998. Phenotypic plasticity in aphids *Myzus persicae* in response to plant secondary chemicals, 189-195. In Nieto Nafria, J. M. and A. F. G. Dixon. [eds.]. Aphids in natural and managed ecosystems. Universidad de Leon, Spain.

Görür, G. 2000. The role of phenotypic plasticity in host race formation and sympatric speciation in phytophagous insects, particularly in aphids. Turkish Journal of Zoology, 24: 63-68.

Groman, J. D. and O. Pelylmyr. 2000. Rapid evolution and specialization following host colonization in yucca moth. Journal of Evolutionary Biology, 13: 223-236.

Guldemond, J. A. 1990a. On aphids, their host plants and speciation. PhD Thesis, Wageningen, University, The Netherlands.

Guldemond, J. A. 1990b. Evolutionary genetics of the *Cryptomyzus*, with a preliminary analysis of the inheritance of host plant prefence, reproductive performance and host alternation. Entomologia Experimentalis et Applicata, 57: 65-76.

Hawthorne, D. J. and S. Via. 2001. Genetic linkage of ecological specilization and reproductive isolation in pea aphids. Nature, 412: 904-907.

Heitland, W. and H. Pschorn-Walcher. 1992. Biological differences between populations of *Platycampus luridiventris* feeding on different species of alder. Entomology Genetics, 17: 185-194.

Holloway, G. J. and S. Povey, and R. Sibly. 1990. The effect of new environment on adapted genetic architecture . Heredity, 64: 323-330.

Hunter, M. D. And J. N. McNeil. 2000. Geographic and parental influences on diapause by a polyphagous insect herbivore. Agricultural and Forest Entomology, 2: 49-55.

Huxley, T. H. 1858. On the agamic reproduction and morphology of Aphis-Part 1. Transactions of the Linnean Society, 22: 193-219.

Jaenike, J. 1981. Critera for ascertaining the existence of host races. American Naturalist, 117: 830-834.

Jorg, E. and Lampel, G. 1996. Enzyme electrophoretic studies on the *Aphis fabae* group. Journal of Applied Entomology, 120: 7-18.

Kaneko, K. 2002. Symbiotic sympatric speciation: consequence of interaction-driven phenotype differentiation through developmental plasticity. Population Ecology, 44: 71-85.

Kennedy, C. E. J. 1986. Attachment may be basis for specialization in oak aphids. Ecological Entomology, 11: 291-300.

Komazaki, S. 1986. The inheritance of egg hatch timing of the overwintering egg among populations of *Aphis citricola* Van der Groot (Homoptera: Aphididae) on the two winter hosts. Kontyü, 54: 48-53.

Kreslavsky, A. G. and A. V. Mikheev, V. M. Solomatin, and V. V. Gritzenko. 1981. Genetic exchange and isolating mechanisms in sympatric races *Lochmaea caprea* (Coleoptera: Chrysomelidae). Zoologicheskij Zhurnal, 60: 62-68.

Leclaire, M. and R. Brandl. 1994. Phenotypic plasticity and nutrition in a phytophagous insect:consequences of colonizing a new host. Oecologia, 100: 379-385.

Mackenzie, A. 1996. A trade-off for host plant utilization in the black bean aphid, *Aphis fabae*. Evolution, 50(1): 155-162.

Mackenzie, A. and J. A. Guldemond. 1994. Sympatric speciation in Aphids II. Host race formation in the face of gene flow. pp: 379-395. *In* S. R Leather and A. D. Watt, N. J. Mills, and K. F. A Walters. [eds.] Individuals, Populations and Patterns in Ecology Andover, Hampshire.

Masutti, F. V. and P. Chavighny. 1998. Host-based genetic differentiation in the aphid *Aphis gossypii* Glover, evidenced from RAPD fingerprints. Molecular Ecology, 7: 905-914.

Maynard Smith. J. 1966. Sympatric speciation. American Naturalist, 100: 637-650.

Mayr, E. 1963. Animal Species and Evolution. Belknap press, Cambridge.

McPheron, B. A. and D. C. Smith, and S. H. Berlocher. 1988. Genetic differences between host races of the apple maggot fly. Nature, 336: 64-66.

Menken, S. B. 1981. Host races and sympatric speciation in small ermine moths, Yponomeutidae. Entomologia Experimentalis et Applicata, 30: 280-292.

Miller, D. R. and M. Kosztarab. 1979. Recent advances in the study of scale insects. Annual Review of Entomology, 24: 1-27.

Mopper, S. 1996. Adaptive genetic structure in phytophagous insect populations. Trends in Ecology and Evolution, 11: 235-238.

Moran, N. A. 1986. Morphological adaptations to host plants in *Uroleucon* (Homoptera: Aphididae). Evolution, 40: 1044-1050.

Moran, N. A. 1987. Evolutionary determinants of host specificity in Uroleucon. pp.29-38 *In* Holman, J. and J. Pelikan, A. F. G. Dixon and L. Weismann. [eds.] Population structure, genetics and taxonomy of aphids and tysanoptera. SPB Academic Publishing, The Hague, Netherlands.

Müller, F. P. 1982. Das problem *Aphis fabae*. Zeitschrift fuer Angewandte Entomologie, 94: 432-446.

Müller, F. P. 1985. Genetic and evolutionary aspects of host choice in phytophagous insects:especially in aphids. Biologiesches Zentralblatt, 104: 225-237.

Muralirangan, M.C., M. Muralirangan, and P.D. Partho. 1997. Feeding behavior and host selection strategies in acridids. Pp. 163-182 *In* Gangwere, S.K., M.C. Muralirangan, and M. Muralirangan [eds] The Bionomics of Grasshoppers, Katydids and their Kin. CAB International, Wallingford, Oxon, UK.

Nason, J. D. and S. B. Heard, and F. R. Williams. 2002. Host-associated genetic differentiation in the goldenrod elliptical-gall moth, *Gnorimoschema gallaesolidaginis* (Lepidoptera : Gelechidae). Evolution, 56(7): 1475-1488.

Nice, C. C. and A. M. Shapiro. 2001. Population Genetic Evidence of Restricted Gene Flow Between Host Races in the Butterfly Genus *Mitoura* (Lepidoptera: Lycaenidae) Annals of the Entomological Society of America, 94(2): 257-267.

Nylin, S. 1988. Host plant specialization and seasonality in a polyphagous butterfly, *Polygonia c-album* (Nymphalidae). Oikos, 53:381-386.

Pankiw, T. and M. L. Winston, M. K. Fondrk, and K. N. Slessor. 2000. Selection on worker honeybee responses to queen pheromone (*Apis mellifera* L.). Naturwissenchaften, 87: 487-490.

Peppe, F. J. and C. Lomonaco. 2003. Phenotypic plasticity of *Myzus persicae* (Hemiptera: Aphididae) raised on *Brassicae oleracea* L. var. acephala (kale) and *Raphanus sativus* L. (radish). Genetics and Molecular Biology, 26(2): 189-194.

Pike, N. and C. Braendle, and W. A. Foster. 2004. Seasonal extension of the soldier instar as a route to increased defence investment in the social aphid *Pemphigus spyrothecae*. Ecological Entomology, 29: 89-95.

Pratt, G. and T. K. Wood. 1992. A phylogenetic analysis of the *Enchenopa binotata* species complex using nymphal characters. Systematic Entomology, 17: 351-357.

Price, P. W. and M. F. Wilson. 1976. Some consequences for a parasitic herbivore, the milweed longhorn beetle, *Tetraopes tetrophthalmus*, of a host plant shift from *Asclepias syriaca* to *A. verticillata*. Oecologia, 25: 331-340.

Prokopy, R. J. and E. W. Bennett and G. L. Bush. 1971. Mating behavior in *Rhagoletis pomonella* (Diptera: Tephritidae). I. Site of assembly. Canadian Entomologist, 103: 1405–1409.

Prokopy, R. J. and E. W. Bennett and G. L. Bush. 1972. Mating behavior in *Rhagoletis pomonella* (Diptera: Tephritidae). I. Temporal organization. Canadian Entomologist, 104: 97–104.

Prokopy, R. J. and S. R. Diehl, and S. S. Cooley. 1988. Behavioral evidence for host races in *Rhagoletis pomonella* flies. Oecologia, 16: 138-147.

Raijmann, L. E. L. and S. B. J. Menken. 2000. Temporal variation in the genetic structure of host associated populations of the small ermine moth *Yponomeuta padellus* (Lepidoptera : Yponomeutidae). Biological Journal of Linnean Society, 70: 555-570.

Rice, W. R. 1987. Speciation via habitat specialization: the evolution of reproductive isolation as a correlated character. Evolutionary Ecology, 1; 301-314.

Roessingh, P., A. Bouaichi, and S.J. Simpson. 1998. Effects of sensory stimuli on the behavioural phase state of the desert locust, Schistocerca gregaria. Journal of Insect Physiology, 44: 883-893.

Roff, D. A. and M. J. Bradford. 2000. A quantitative genetic analysis of phenotypic plasticity of diapause induction in the cricket Allonemobius socius. Heredity, 193-200.

Rogers, S. M. and S. J. Simpson. 1997. Experience-dependent changes in the number of chemosensory sensill on the mouthparts and antennae of Locusta migratoria. Journal of Experimental Biology, 200: 2313-2321.

Rossi, A. M. and P. Stiling, M. V. Cattell, and T. I. Bowdish. 1999. Evidence for host-associated races in a gall-forming midge: trade offs in potential fecundity. Ecological Entomology, 24: 95-102.

Sandström, J. 1996. Temporal changes in host adaptation in the pea aphid, Acyrthosiphon pisum. Ecological Entomology, 21: 56-62.

Sakata, K. and Y. Ito, J. Yukawa, and S. Yamane. 1991. Ratio of sterile soldiers in the bamboo aphid, Pseudoregma bambucicola (Homoptera: Pemphigidae), colonies in relation to social and habitat conditions. Applied Entomological Zoology, 26: 463-468.

Sattman, D. A. and R. B. Cocroft. 2003. Phenotypic plasticity and repeatability in the mating signals of Enchenopa treehoppers, with implications for reduced gene flow among host-shifted populations. Ethology, 109: 981-994.

Scheiner, S. M. 1998. The genetics of phenotypic plasticity . VII. Evolution in a spatially structured environment. Journal of Evolutionary Biology, 11: 303-320.

Schlichting, C. D. and M. Pigliucci. 1998. Phenotypic evolution. Sinauer, Sunderland, MA.

Schültze, M. and U. Maschwitz. 1991. Enemy recognition and defence within trophobiotic associations with ants and by the soldier caste of Pseudoregma sundanica (Homoptera: Aphidoidea). Entomologia Generalis, 16: 1-12.

Schwarz, D. and B. A. McPheron, G. B. Hartl, E. F. Boller, and T. S. Hoffmeister. 2003. A second case of genetic host races in Rhagoletis? A population genetic comparison of sympatric host populations in the European cherry fruit fly, Rhagoletis cerasi. Entomologia Experimentalis et Applicata 108: 11-17.

Sembene, M. and A. Delobel. 1998. Genetic differentiation of groundnut seed-beetle populations in Senegal. Annual Review of Entomology, 41: 325-352.

Sezer, M, and R. K. Butlin. 1998a. The genetic basis of host plant adaptation in the brown planthopper (Nilaparvata lugens). Heredity, 80: 499-508.

Sezer, M, and R. K. Butlin. 1998b. The genetic basis of oviposition preference differences between sympatric host races of the brown planthopper (Nilaparvata lugens). Philosophical Transactions of the Royal Society of London. Series B, 265: 2399-2405.

Shapiro, A. M. 1978. The evolutionary significance of redundancy and variability in phenotypic-induction mechanisms of pierid butterflies (Lepidoptera). Psyche, 85: 275-283.

Shaposhnikov, G.Kh. 1965. Morphological divergence and convergence in an experiment with aphids (Homoptera, Aphidinea). Entomological Review, 44:1-12.

Shaposhnikov, G.Kh. 1966. Origin and breakdown of reproductive isolation and the criterion of the species. Entomological Review of Wash, 45: 1-18.

Shibao, H. and J. M. Lee, M. Kutsukake, and T. Fukatsu. 2003. Aphid soldier differentiation: density acts on both embryos and new born nymphs. Naturwissenchaften, 90: 501-504.

Shufran, K. A. and J. D. Burd, and G. Lushai. 2000. Mitochondrial DNA sequence divergence among greenbug (Homoptera: Aphididae) biotypes: evidence for host adapted races. Insect Molecular Biology, 9: 179-184.

Simpson, S.J., E. Despland, B.F. Hagele, and T. Dodgson. 2001. Gregarious behavior in desert locusts is evoked by touching their back legs. Proceedings of the National Academy of Sciences of the United States of America, 98: 3895-3897.

Southwood, T. R. E. 1986. Plant surfaces and insects- an overview, 1-22. *In* Juniper, B. And T. R. E. Southwood, [eds.]. Insects and the plants surface. London, Edward Arnold.

Stern, D.L. and W. A. Foster. 1996. The evolution of soldiers in aphids. Biological Review, 71:27-79.

Strong, D. R. and D. S. Simbreloff, L. G. Abele, and A. B. Thistle. 1984. Ecological, communities: conceptual issues and and the evidence. Princeton University Press.

Sultan, S.E. and H. G. Spencer. 2002. Metapopulation structure favors plasticity over local adaptation. American Naturalist, 160: 271-283.

Sunose, T. and S. Yamane, K. Tsuda, and K. Takasu. 1991. What do the soldiers of *Pseudoregma bambucicola* defend? Japanese Journal of Entomology, 59: 141-148.

Tanaka, S. and Y. Ito. 1994. Interrelationships between the eusocial aphid, *Pseudoregma bambicicola*, and its syrphid predator, *Eupeodes confrater*. Japanese Journal of Entomology, 63: 221-228.

Tauber, M. J., C. A. Tauber, and S. Masaki. 1986. Seasonal Adaptations of Insects. Oxford University Press, New York.

Tauber, C. A. and M. J. Tauber. 1989. Sympatric speciation in insects; perception and perspective, 307-344. *In* Otte, D. and A. Endler, [eds.]. Speciation and its consequences.Sianuer Associates, Inc. Sunderland, MA.

Tilmon, K. J. and T. K. Wood, and J. D. Pesek. 1998. Genetic variation in performance traits and the potential for host shifts in *Enchenopa* treehoppers. Annales Entomological Society of America, 91: 397-403.

Thompson, J. D. 1991. Phenotypic plasticity as a component of evolutionary change. Trends in Ecology and Evolution, 6: 246-249.

Thompson, D.B. 1992. Consumption rates and the evolution of diet-induced plasticity in the head morphology of *Melanoplus femurrubrum* (Orthoptera: Acrididae). Oecologia, 89:204-213.

Uvarov, B. 1966. Grasshoppers and Locusts Vol. 1. Cambridge University Press, London. 481 pp.

Uvarov, B. 1977. Grasshoppers and Locusts Vol. 2. Centre for Overseas Pest Research, London. 613 pp.

Vanbergen, A. J. and B. Raymond, I. S. K. Pearce, A. D. Watt, R. S. Hails, and S. E. Hartley.2003. Host shifting by *Operophtera brumata* into novel environments leads to population differentiation in life history traits. Ecological Entomology, 28: 604-612.

Via, S. 1991. The genetic structure of host plant adaptation in spatial patchwork: demographic variability among reciprocally transplanted pea aphid clones. Evolution, 45: 827-857

Via, S. 1999. Reproductive isolation between sympatric host races of pea aphids. I. Gene flow restriction and habitat choice. Evolution, 53: 1446-1457.

Via, S. 2001. Sympatric speciation in animals: the ugly duckling grows up. Trens in Ecology and Evolution, 16(7): 381-390.

Via, S. and R. Lande. 1985. Genotype-environment interaction and the evolution of phenotypic plasticity. Evolution, 39: 505-522.

Via, S. and R. Gomulkiewicz, G. de Jong, S. M. Scheiner, C. D. Schlichting, and P. H. Van Tienderen. 1995. Adaptive phenotypic plasticity: consensus and controversy. Trends in Ecology and Evolution, 10 (5): 212-217.

Walsh, B. D. 1864. On phytophagic varieties and polyphagous species. Proceedings of the Entomological Society of Philadelphia, 3: 403-430.

Waring, G. L. and W. G. Abrahamson, and D. J. Howard. 1990. Genetic differentiation among host–associated populations of the gallmaker · *Eurosta solidaginis* (Diptera: Tephritidae). Evolution, 44: 1648-1655.

Watt, A. D. and A. F. G. Dixon. 1981. The role of cereal growth stages and crowding in the induction alatae in *Sitobion avenae* and its consequences for population growth. Ecological Entomology, 6: 441-447.

Weiblen, G. D. and G. L. Bush. 2002. Speciation in fig pollinators and parasites. Molecular Ecology, 11: 1573-1578.

Weisser, W. W. and B. Stadler. 1994. Phenotypic plasticity and fitness in aphid. European Journal of Entomology, 91: 71-78.

West-Eberhard, M. J. 1989. Phenotypic plasticity and the origins of diversity. Annual Review of Ecology and Systematics, 20: 249-278.

West-Eberhard, M. J. 2003. Developmental plasticity and evolution. Oxford University Press,794 pp.

Whitman, D.W. 1990. Grasshopper chemical communication. pp. 357-391 *In* Chapman, R.F. and A. Joern [eds.] Biology of Grasshoppers. Wiley, New York.

Wilson, E. O. 1992. The Diversity of Life. Belknap Press, Cambridge, MA. 424 pp.

Wood, T. K. and K. J. Tilmon, A. B. Shantz, and C. K. Harris. 1999. The role of host-plant fidelity in initiating insect race formation. Evolutionary Ecology Research, 1:317-332.

Wool, D. and D. F. Hales. 1997. Phenotypic plasticity in Australian Cotton Aphid (Homoptera:Aphididae): Host plant effects on morphological variation. Annual Entomological Society of America, 90: 316-328.

Wright, S. 1931. Evolution in mendelian populations. Genetics, 16: 97-159.

Yang, Y. and A. Joern. 1994. Gut size changes in relation to variable food quality and body size in grasshoppers. Functional Ecology 8: 36-45.

Zhivotovsky, L. A. and M. W. Feldman. and A. Bergman. 1996. On the evolution of phenotypic plasticity in a spatially heterogeneous environment. Evolution, 50: 547-558.

8

Adaptive Allometric Responses of Galling Insects to Availability of Ovipositing Sites

Andréa Lúcia Teixeira de Souza[1], Marcel Okamoto Tanaka[2] and Rogério Parentoni Martins[3]

[1]*Departamento. de Biologia - CCBS - CP 549, Universidade Federal de Mato Grosso do Sul, Campo Grande, MS, 79070-900, Brazil, Email: altsouza@nin. ufms.br*

[2] *Departamento de Biologia - CCBS, Universidade Federal de Mato Grosso do Sul - CP 549, CEP 79070-900, Campo Grande, MS, Brazil, Email: martnk@yahoo.com*

[3] *Laboratório de Ecologia e Comportamento de Insetos – Departamento de Biologia Geral, ICB, Universidade Federal de Minas Gerais, CP 486, CEP 30161-970, Belo Horizonte, MG, Brazil, Email: wasp@icb.ufmg.br*

Introduction

Galls are interesting biological and evolutionary entities because their phenotypes arise from the interaction of the genotypes of two species – the plant and the gall maker. Gall-making insects have an intimate relationship with their host plants; they manipulate the plants' normal tissue development, causing it to form a structure that both shelters and feeds the parasite, at the plant's expense (Weis et al. 1988). The galling insect produces the stimulus, and the host plant determines the developmental response that produces the gall (Abrahamson and Weis 1987).

Plants and gall-makers represent two trophic levels. However, two additional players impact this system: the natural enemies of gall insects (the 3rd trophic level), and other herbivores. Natural enemies can impose strong selective pressures on gall phenotypes or constrain their distribution, since the attack of natural enemies can vary both among host plants and sites (Price and Clancy 1986, Abrahamson and Weis 1987, Hawkins et al. 1997). Likewise, other herbivore species can influence plant-gall insect relationships, when they cause or induce changes in the host plant. Thus, the aspect diversity of galls and their distribution could be an adaptive consequence of the interaction between gallers, host plants, natural enemies, and competitors.

In this chapter we review how differences between and within individuals of a plant species constitute an heterogeneous resource base for ovipositing female gall insects, influencing the survival of galling insects' offspring. We find a great variation on the relationship between the availability of favorable oviposition sites and gall distribution, so that galls can develop on suboptimal sites. Models of plant-gall-insect interactions predict, and empirical studies often indicate, a strong positive correlation between the preference for oviposition sites and the offspring's performance (Whitham 1980, Price et al. 1987a, Craig et al. 1989, Hartley and Lawton 1992, Burstein and Wool 1993, Larsson and Ekbom 1995, Fay et al. 1996, Cronin and Abrahamson 1999). We also find evidence for phenotypic variation in life-history patterns, in traits such as gall size, growth rates and body size of the offspring. We discuss hypotheses on the effects of plants as heterogeneous habitats, and phenotypic plasticity as adaptive to increase the availability of oviposition sites for galling insects.

Response of Host Plants to Gall Induction and Choice of Oviposition Sites

The high specificity of galling insects to their host plants can result in a narrow range of egg-laying options for females searching for adequate oviposition sites. This narrow range can become even narrower when variation between and within host plants are considered (Abrahamson and Weis 1987, Weis 1992). Individual host plants are not homogeneous, due to variation in environmental conditions and to plant genetic variability (Horner and Abrahamson 1992, Stiling and Rossi 1996). Even within the same genotype at a given site, different parts of the same plant may differ in their susceptibility to attack and induction by the galling insect (Whitham et al. 1984, Price 1991, Burstein and Wool 1993). Female gall insects should select host plants or parts of plants in which growth, survival, and reproduction of their offspring are maximized (Thompson 1988). This constraint is tighter for galling insects because established larvae cannot move to higher quality sites (Walton et al. 1990). Nevertheless, a preference-performance relationship is not always found for gall-making species, either due to a lack of preference or because their performance is uncorrelated with egg-laying characteristics of the chosen sites (Price et al. 1999, Fritz et al. 2000, Kokkonen 2000, Cronin and Abrahamson 2001, but see Craig et al.

1989, Horner and Abrahamson 1999, Pires and Price 2000). Differences among and within plants constrain the availability of oviposition sites and influence the development of galls, resulting in variation in the performance of the offspring within a plant population.

Variation Between Plants

Several factors influence individual plant responses to galling insects. Differences in developmental stages of plants, environmental conditions, mechanisms of plant defense, amongst others, influence egg-laying behavior and success of gallers. Plant age and phenological stage determine the magnitude of plant response. For example, the budburst stage of the host plant can be critical for the establishment of galls, and for several systems the timing of adult hatching and oviposition are well synchronized with the timing of bud formation and growth (Junichi 2000, Shibata 2001, Wool 2004). However, within a population the phenological stages of individual plants may vary, and the maintanance of variable hatching times within a population of gallers could enable a better usage of these resources, although there may be limits for this variability (Akimoto and Yamaguchi 1994). Akimoto (1998) found great variation in hatching times within half-sib families of the aphid *Tetraneura* sp. galling on *Ulmus davidiana*, whereas the success in galling of transplanted individuals strongly depended on the synchrony between nymphal hatching and budburst, which varied among individual plants. On the other hand, Rehill and Schultz (2002) found that eggs of the aphid *Hormaphis hamamelidis* hatched in advance of budburst of its host plant, *Hamamelis virginiana*, thus ensuring gall formation.

Plants in distinct developmental stages can present differences in morphology, nutrient availability, and anti-herbivore responses, thus influencing their susceptibility to attack by gallers (Kearsley and Whitham 1989, Price 1991). Kearsley and Whitham (1989) found that the gall-forming aphid *Pemphigus betae* rapidly increased in older clones of narrowleaf cottonwood *Populus angustifolia*, with densities 70 times higher on leaves of old clones when compared to younger ones. Their performance also benefited from a higher availability of oviposition sites (due to a higher frequency of older plants), more favourable resources and less resistance in mature clones. On the other hand, Price et al. (1987b) found a negative correlation of densities of the bud galler *Euura mucronata* and age of clones of *Salix cinerea*, with the younger and more vigorous plants being more attacked. Walton et al. (1990) found that ovipunctured clones of *Solidago altissima* by the stem galler *Eurosta solidaginis* were taller than unpunctured ones, although rates of gall formation in this range of plant heights were similar.

Different individuals of the same plant species do not form a genetically homogeneous population, and some individuals can be more susceptible to galler attack (Anderson et al. 1989). These differences have been used to select herbivore-resistant varieties of economically important species, such as black currant *Ribes nigrum*, attacked by the leaf galler midge *Dasineura tetensi* (Keep 1985, Hellqvist and Larsson 1998, Hellqvist 2001); 6 of 11 varieties tested were susceptible to attack, whereas, 5 showed variable degrees of resistance, mainly by antibiosis (Hellqvist and Larsson 1998). Females of *D. tetensi* did not discriminate among genotypes, and similar oviposition rates were found on all varieties, so the lower development of galls on resistant varieties could be due to inducible defenses against herbivory (Hellqvist and Larsson 1998, Ollerstam et al. 2002). Inducible defenses have recently been recognized as important factors influencing failure to induce galls (Fernandes 1990, Fernandes and Negreiros 2001). Lack of preference for host plant individuals during oviposition and the resulting poor performance of the offspring on resistant genotypes was noted for other species such as *Dasineura marginemtorquens* on *Salix viminalis* (Larsson and Strong 1992, Larsson et al. 1995) and *Eurosta solidaginis* on *Solidago altissima* (Horner and Abrahamson 1992).

Galling species can also discriminate among hybrids of host plant species (Aguilar and Boecklen 1992, Fritz et al. 1994), and even among subspecies within a host plant (Floate et al. 1996). These differences could be related to environmental differences, such as found by Graham et al. (2001) in a transplant experiment of hybrids and subspecies of big sagebrush *Artemisia tridentata* to three distinct environments. They found significant genotype x environment interactions on gall densities, but considered environmental effects stronger than genotype effects because gall density increased with degree of plant stress (Graham et al. 2001). In some cases, plant responses may have a stronger influence. Kokkonen (2000) found no discrimination of two *Pontania* sawfly species between pure and hybrid individuals of *Salix caprea-starkeana*, although one of the species galling on hybrids had higher abortion rates and produced smaller adults.

Plant resistance to gall induction is mainly achieved by chemical defenses (Abrahamson et al. 2003), but other mechanisms of defense can be also found. Williams and Whitham (1986) found premature leaf abscission by two species of cottonwoods, with increased responses at higher densities of *Pemphigus* gall aphids, resulting in mortalities of more than 98 percent of insects on these leaves. Larson and Whitham (1997) showed that susceptibility to the leaf galler *Pemphigus betae* among hybrid clones of cottonwood, *Populus angistifolia* and *P. fremontii*, were related to architectural differences between individuals. Plants with higher densities of buds were less vulnerable to attack by *P. betae*. Resistant plants become more vulnerable when their bud densities were artificially reduced, suggesting that resistant plants inhibited gall development by low availability of photoassimilates and nutrients, maintained through a high density of sinks (buds) (Larson and Whitham 1991, 1997).

Plant resistance and attractiveness can also be influenced by previous attacks of other herbivores, through induced defenses or stunting growth (Cronin and Abrahamson 1999, 2001), thus influencing preference-performance relationships. Cronin and Abrahamson (1999) found that attack by meadow spittlebugs on the goldenrod *Solidago altissima* resulted in a decline in ramet growth rates, reducing host-plant preference for the stem galler *Eurosta solidaginis*. In another experiment, Cronin and Abrahamson (2001) found a decreased preference of *E. solidaginis* for previously attacked ramets of goldenrod by two herbivore species. In both experiments, performance was weakly affected by previous herbivore attack, and was more influenced by differences among goldenrod genotypes.

Moisture and soil nutrients also influence plant physiology, stress, and health, and thus susceptibility to attack by insect gallers. Preszler and Price (1988) found greater oviposition by the shoot-galling sawfly *Euura lasiolepis* on plants of *Salix lasiolepis* not limited by water, whereas larval mortality increased fourfold on drought-stressed plants relative to controls. Björkman (1998) found higher infestation and gall density of the aphid *Sacchiphantes abietis* on Norway spruce plants stressed by drought, when compared to plants subject to different levels of water and nutrient stress, whereas gall size was indirectly influenced by these factors due to differences in the growth rates of the shoots.

These factors do not necessarily influence plant responses directly, and interactions between factors are commonly found. Most studies focused on the interaction between plant genotypes and environment. For example, in conditions of drought stress, Björkman (2000) found higher gall densities and survival of stem mothers of the aphid *Adelges abietis* on resistant Norway spruce individuals, whereas the opposite was found on susceptible individuals. The interaction seems due to a reduced production of a phenolic defense compound by resistant plants, and higher production by susceptible plants in stressed conditions (Björkman 2000). Interactions between genotype, nutrient and shading levels influenced the number of *Solidago altissima* plants ovipunctured by *Eurosta solidaginis*, but larval survival depended only on plant genotype (Horner and Abrahamson 1992). In another experiment, there was no interaction between *S. altissima* genotypes and two watering regimes on oviposition preference by *E. solidaginis*: a significant interaction on the number of galls formed was found, but there were no differences on the proportion of adults emerging (Horner and Abrahamson 1999). Using reciprocal transplants, Stiling and Rossi (1996) found that abundance, gall size and abortion rates of the gall midge *Asphondylia borrichiae* on a coastal plant *Borrichia frutescens* depended both on clone genotype and environment, mainly due to large salinity and shading differences among environments. These experiments suggest that the environment has a strong effect on numerous factors, and that the interaction is probably stronger in harsher conditions (Stiling and Rossi 1996, Björkman 2000). Thus, the response of plants to gall induction may vary both at large and small spatial scales, and variation in these effects can also be found within plants.

Variation Within Individual Plants

Plants have a modular architecture, and several factors restricting the sites of oviposition discussed above may also apply to parts of the same plant (Whitham et al. 1984). Both environmental effects at small spatial scales and plant architecture can lead to differences in growth rates among modules on distinct positions in the plant, which in turn may influence their response to induction by the galler (Price 1991). The plant vigour hypothesis predicts that herbivores should select larger and more vigorous plants or plant modules, resulting in a better performance of their offspring at these sites (Price 1991). Several studies support this hypothesis in gall systems (Kimberling et al. 1990, Woods et al. 1996, Prado and Vieira 1999). In other studies, either no preference was detected (Eliason and Potter 2000, Cornelissen and Fernandes 2001), or performance was similar among modules of differing sizes (Fernandes 1998, Fritz et al. 2000, Rehill and Schultz 2001) or even better at intermediate sizes (McKinnon et al. 1999), and other factors influenced the performance of offspring, such as plant defenses, gall relative position on the plant, gall position in relation to other galls, and manipulation of resources of the host plant by the insect galler.

The position of the gall on the plant, such as in more external or apical positions, could influence their mortality patterns by exposing the gall to different environmental conditions or to more frequent attacks by natural enemies (Confer and Paicos 1985, Tscharntke 1992). Distribution of galls among and within modules can influence also the performance of the offspring through differential use and interception of photoassimilates (Whitham 1980, Inbar et al. 1995, Larson and Whitham 1991, 1997). Stem mothers from galls of *Pemphigus betae* located in the distal position of leaves of *Populus angustifolia* had almost one-third the mass, one-fifth the progeny, and 18 times the failure rate of those in the basal position (Whitham 1980). Inbar et al. (1995) found evidence of interspecific competition between the aphids *Geoica* sp. and *Forda formicaria* galling on leaflets of *Pistacia palaestina*. Both species are sinks for plant assimilates, but *Geoica* sp. is a stronger sink, causing senescence of the leaflet and limiting the resources available for *F. formicaria*. When raised in competition with *Geoica*, *F. formicaria* suffered 84% mortality and 20% reduction in reproductive success (Inbar et al. 1995). As discussed above, competition with sinks from plant organs can reduce the performance of the gallers (Larson and Whitham 1997).

Since galls may act as sinks of photoassimilates, higher densities of galls within a module could be benefitial for gall development by increasing the strength of the sink (Craig et al. 1990). In this way, terminal buds can be destroyed, and dorment buds develop, so that the availability of adequate (young) branches for gall development is maintained (Craig et al. 1986, 1990), although excessive attack rates can be detrimental to the galler (Prado and Vieira 1999). The manipulation of host-plant resources can also reduce the effects of the environment, by maintaining optimal resource levels for larval survival. The survivorship of the cynipid galler *Neuroterus quercus-baccarum* was negatively correlated with nitrogen levels, but nutrient levels in oak (*Quercus robur*) gall tissues were unaffected by experimental nitrogen addition (Hartley and Lawton 1992).

Characteristics of host plants and the environment can strongly influence both preference for oviposition sites and the performance of the resulting offspring. Females do not always choose the best oviposition sites for offspring growth and survival, but adaptations such as host-plant manipulation and phenotypic plasticity of gall and larval traits could increase the chances of galler survival and reproductive success (Weis 1992).

The Relationship between Phenotypic Plasticity of Gall-Makers and their Distribution

Several studies on insect gallers recorded a great variation in life-history traits and patterns, such as clutch size, gall formation rates, gall size, number of individuals per gall, developmental times, and sizes of larvae, pupae, and adults (Whitham 1978, Weis and Gorman 1990, Weis et al. 1992, Akimoto and Yamaguchi 1994, Freese and Zwölfer 1996, Desouhant et al. 1998, Rossi et al. 1999, Sopow and Quiring 2001). These characteristics are generally correlated with each other; for example, gall size may depend on the number of individuals per gall, whereas longer developmental times normally result in larger adults (Freese and Zwölfer 1996, Björkman 2000, Souza et al. 2001). However, this variation is not always adequately explored or explained in those studies, and it is assumed as a natural variation. We argue that, at least in some cases, this variation can be adaptive, and constitutes a phenotypic plasticity that enables the use of extremely heterogeneous plant resources.

Plasticity in developmental times and larval body sizes could be an adaptive strategy, because the availability of favourable oviposition sites varies due to numerous factors, including environmental stochasticity. When high-quality habitats are few, the ability to form galls and grow in low-quality habitats could allow survival, even though resulting body sizes and offspring fecundities are smaller. Tolerance for low quality habitats could likewise increase the geographical distribution of these species.

Growth rates are generally assumed to be maximized within the limits imposed by the environment. The developmental time influences body size (Nylin and Gothard 1998), which influences age of first reproduction, fecundity, and survival rates (Roff 1992, Stearns 1992, Honek 1993, Nylin and Gothard 1998, but see Klingenberg and Spence 1997). Rates of offspring survival, growth, and fecundity have been used as performance indicators of galling insects in studies on their spatio-temporal distribution, through population dynamics and female ovipositing behaviour (Walton et al. 1990, Craig et al. 1999, Horner and Abrahamson 1999, Horner et al. 1999, Cronin and Abrahamson 2001).

Few studies have directly addressed the norm of reaction for distinct characteristics of gallers, and the factors involved in such variation. One of the best examples is the shoot galler *Eurosta solidaginis* on *Solidago altissima* (Weis and Gorman 1990, Weis et al. 1992, Weis and Kapelinski 1994, Weis 1996). *E. solidaginis* forms variable gall sizes and, although gall sizes are not directly related to adult sizes and fecundities (Weis and Abrahamson 1986, Lichter et al. 1990), there is a size-differential survival, due to selection pressures exerted by its natural enemies. Preferential attack by parasitoids on small galls would select for larger gall sizes, whereas attack by predator birds would have an opposite effect (Abrahamson et al. 1989, Weis and Abrahamson 1986). The fitness function relative to gall size would have higher values on intermediate gall sizes; however, the general effects of natural enemies is variable among populations, and in many cases the effect of parasitoids is larger than those of predators, resulting in a net upward selection on gall size (Weis et al. 1992, Weis and Kapelinski 1994). Depending on the environmental condition, plant responses could also vary among populations, so that local intensity of selection could be strongly dependent of the mean and variance of gall sizes (Weis et al. 1992). For example, the effects of parasitoids would be stronger in a population with smaller mean gall sizes than in a population with bigger mean gall sizes. Thus, even being an heritable trait, variation in gall size could be influenced not only by the effects of natural enemies, but also on spatial and temporal variation of the environment, resulting in variable availability of resources for the plant and, by extension, to the galler (Weis 1996).

The phenotypic variation found in galling insects is not always correlated with the amount of resources found in plant organs such as leaves and shoots. Akimoto and Yamaguchi (1994) showed that the performance of the galling aphid *Tetraneura* sp. on *Ulmus davidiana* was more related to synchronization with timing of bud formation, with early hatching fundatrices (which reproduce parthenogenetically in the holocycle - see Wool 2004) having larger fecundities by using early developing leaves that were better quality sites than leaves developing later, independently of its size and position on the host tree. Galls that developed later had higher mortality and lower growth rates (Akimoto 1998). However, there was strong selection against early hatching larvae due to mortality caused by environmental stress, resulting in stabilizing selection. Variation in hatching times would also reduce competitive interactions because the timing of bud formation varies among and within trees. In this way, contrary to other aphid species, restriction for oviposition sites is not so tight, and no territorial behavior was observed in this species (Whitham 1978). Continuous variation in hatching times thus favoured the performance of *Tetraneura* sp. by reducing competition for high quality sites, which were also continuously provided along the plant´s growing season (Akimoto 1998). If no plasticity for this trait was observed, only part of the available oviposition sites created during the growing season of its host plant would be used, possibly reducing the performance of the gallers.

Variation in the abundance and performance of galling insects is often related to plant age (Price et al. 1987b, Walton et al. 1990, Woods et al. 1996) and vigour, the latter generally being measured as leaf or shoot sizes or growth rates (Price 1991, Glynn and Larsson 1994, De Bruyn 1994, Fay et al.

1996, Price et al. 1997, Björkman 1998, Ozaki 2000). McKinnon et al. (1999) evaluated the effect of variable foliar chemical composition on shoots of different sizes of white spruce *Picea glauca* on the abundance and growth rates of the shoot galler aphid *Adelges abietes*. They carried out an experiment to create a gradient of shoot sizes by using six treatments receiving different combinations of nutrients and root pruning. Their results showed a significant effect of shoot size on gall abundance, with a higher number of galls being formed on intermediate sized shoots. Further, the number of *A. abietes* individuals was positively correlated with gall volume. As female fecundity is positively correlated with adult body size, they suggested that the performance of the offspring depended on the choice of oviposition sites by the females relative to the shoot size where eggs were laid. The relationship found by McKinnon et al. (1999) between *A. abietes* adult females and gall volume was very low (r^2=0.12) but, according to their results, since gall volume depended on the number of individuals it contained, it seems that the effect of gall volume on female adult sizes was confounded with the effect of the number of individuals inside the gall. Thus, the amount of resources available for each individual could not be evaluated, since in large galls the resources are shared by many individuals (see below).

Several studies found that gall sizes are directly related to the number of individuals inside the galls (Raman and Abrahamson 1995, Wool and Bar-El 1995, Austin and Dangerfield 1998, Ozaki 2000). However, in many of these studies variation in gall sizes seems to depend on the plant response to gall induction (Wool and Manheim 1988, Clancy et al. 1993, Raman and Abrahamson 1995). If the plant response determines only the proportion of the number of eggs that begin their development, we would expect a perfect fit of a model considering just gall volume as a function of the number of individuals within the gall. However, this fit is not so good in many of the studies (Ozaki 1993, 2000, Austin and Dangerfield 1998, Mckinnon et al. 1999, Sopow and Quiring 2001). Thus, could galls with similar number of individuals have different volumes? If so, large galls in relation to the number of individuals should have a greater availability of resources per insect, resulting in greater performance of gall makers. In some species of galling insects, gall volume is only related with the number of individuals enclosed, and thus smaller galls correlate with smaller clutch size or higher mortality, although in other systems galls with the same number of individuals but with different volumes can be formed, caused by variation in plant responses. Further, some species have very flexible ranges of body sizes, and the low availability of resources found in small galls results in a reduction of offspring body sizes. Thus, the proportion of individuals that die when developing in low quality habitats should vary with the ability of the individuals to get smaller, e.g., with higher plasticity in body sizes.

Plasticity of body sizes must vary among species of galling insects, although this is rarely mentioned. For example, Sopow and Quiring (2001) evaluated the relationship between adult body sizes and gall volume of three species of spruce-galling aphids, *Adelges cooleyi*, *Pineus pinifoliae* and *P. similis*. They grew ~70 galls of each species growing in different conditions (e.g., in different plant species and heights on the plants). The relationship between gallicolae mean sizes – as expressed by the mean length of the wings of hatched adults – was significant for *P. pinifoliae* only. These results showed that gall volume determines only the number of individuals, but not body size for both *A. cooleyi* and *P. similis*. A linear model was adjusted to evaluate the effect of gall density – expressed as the number of individuals per unit of gall volume – on the body sizes of gallicolae hatching from these galls. The results of Sopow and Quiring (2001) showed a significant relation for *P. pinifoliae* (r^2 = 0.47, p<0.001) but only a weak one for *P. similis* (r^2 = 0.08, p=0.015). Thus, it seems that *P. pinifoliae* has a more plastic body size when compared to the other two species, adjusting body size to the amount of resources per individual. For another aphid species, *Adelgis japonicus*, Ozaki (1993, 2000) found a strong positive relationship between gall volume and the number of individuals enclosed, suggesting that some of the offspring do not survive when in excess. Thus, the ability to adjust body size in response to resource availability varies greatly even among related species, as in the case of *Pineus*.

Genetic variation among individuals or populations of host plants can result in differences of plant resistance to gall induction, and the insects can fail to develop in a gall (see above). Some plant species can have more resistant or susceptible genotypes relative to gall formation than others (Horner and Abrahamson 1992, 1999, Hellqvist and Larsson 1998, Rossi et al. 1999). On the other hand, some genotypes within populations of galling insects can induce galls on resistant plants (Hellqvist 2001).

The eventual colonization of related host plant species can result in reproductive isolation through the formation of sympatric host races (reviewed by Abrahamson et al. 2001). Abrahamson and coworkers found that the gall fly *Eurosta solidaginis* attacks *Solidago altissima* in most of its geographical range, but in Northern USA and Southern Canada *E. solidaginis* can also induce galls on *S. gigantea*, maintaining different races through behavioral reproductive isolation (Craig et al. 2001, Abrahamson et al. 2001). However, this process may not be widespread to galling insects due to their high specificity in their host plants, as well as distinct sets of anti-herbivore defenses to cope with (Abrahamson et al. 2003). Nevertheless, the existence of different phenotypes that can colonize new species of host plants due to a lower discrimination during oviposition by the herbivore insects, combined with their ability to develop in low quality patches enable the exploration of a resource less subject to competition and attack by natural enemies (Brown et al. 1995, Craig et al. 2000).

Rossi et al. (1999) evaluated differences in the performance of the gall midge *Asphondylia borrichiae* within and among three host plant species, *Borrichia frutescens*, *Iva frutescens* and *I. imbricata*. They evaluated pupal body sizes and survival rates of the galling insect as indicators of the performance of the offspring. Gall sizes of *A. borrichiae* differed among the three plant species, whereas the number of chambers was constant. Gall density (= the number of chambers per cm^3 of gall) differed significantly among the plant species, with *I. frutescens* > *I. imbricata* > *B. frutescens*. As a result, gall densities negatively influenced pupal weight. Even within plant species, they detected a negative relationship between both variables. Pupal weight varied from ~ 0.5 to 2.2 mg in *B. frutescens*, 0.7 to 1.6 mg in *I. imbricata* and 0.3 to 1.7 mg in *I. frutescens*. The developmental time of *A. borrichiae* also varied among the three plant species: it was about two times smaller in *I. imbricata* (48 days) – where gall densities were higher – than on *B. frutescens* (106 days). Rossi et al. (1999) concluded that *A. borrichiae* were evolving toward the formation of different races: if the level of parasitism on the galler is higher on plants more susceptible to it, the chances of survival on less susceptible host plants increase, although with a lower level of parasitism, therefore being advantageous for *A. borrichiae*. Their results indicate that survival on non-host plant species is costly, resulting in a reduction of the body sizes of the offspring. They suggest that escape from natural enemies could have exerted a strong selection pressure for the midge to switch among host plants. However, this was possible only because this species already had a high flexibility to establish itself in low quality habitats of its original host plant by adjusting its body size to the availability of resources.

One of the most complete studies on variation of life-history traits of galling insects was carried out by Freese and Zwölfer (1996). They evaluated how variation in brood sizes of the tephritid fly *Urophora cardui*, that induces galls on shoots of creeping thistle *Cirsium arvense*, influenced offspring performance. Females oviposited a variable number of eggs per clutch (1-29), averaging 11.9 eggs, but the success in the development of the chambers was low, with about four chambers formed in populations of Central Europe. The increase in gall size resulted in an increase of individual mean weights, with large variations on body weights of larvae (~ 5-25 mg), pupae (~ 4-20 mg), and adults (~ 1-15 mg), and higher survival and fecundity of the larger individuals. On the other hand, gall size varied both with the number of eggs laid (clutch size) and plant response. Thus, the production of large galls would be advantageous for female fitness, resulting in higher larval survival rates, increased weights of adults and consequently increased number of eggs left for the next generation. However, most galls of *U. cardui* were below the optimum size and were unfavorably small, and the authors suggest that few chambers develop under unfavorable conditions, resulting in smaller galls. Gall size and duration of developmental times are strongly influenced by environmental conditions and quality of the host plant, particularly by availability of nutrients for the plants and drought stress. Unfortunately, the authors did not mention whether galls of variable sizes can be formed for a given number of chambers. If these differences exist, could the increase in density within the gall (reduced gall size in relation to the number of chambers) influence negatively adult body sizes in high-density conditions? This study showed a huge variation in the body sizes of adult individuals surviving even under unfavorable conditions, suggesting that the phenotypic plasticity of this species can allow the utilization of sub-optimal sites.

A Case Study

We explored how plasticity in reproductive traits and adult size influence the distribution of the shoot galler weevil *Collabismus clitellae* Boheman (Coleoptera: Curculionidae) on the "lobeira," *Solanum lycocarpum* St. Hil. (Solanaceae), in Southeastern Brazil. In this section, we review the published data on this weevil-host plant system and present new evidence on the adaptive value of those variations. We conclude that plasticity in life-history traits allows this species to increase its host range.

Solanum lycocarpum is a common woody shrub in central and Southeastern Brazil, growing to a maximum of 5 m in height. It occurs frequently in disturbed areas such as pastures, grasslands, and roadsides, where plants are frequently pruned resulting in large patches of fast-growing shoots in young shrubs (1-2 years old) smaller than 1.5 m in height (Carvalho 1985). These shrubs host *C. clitellae*, which can reach high numbers in the large patches of *S. lycocarpum*.

Female *C. clitellae* lay eggs in early summer (November and December), by chewing ovipositing holes in young, non-woody stems of the host plant

(Souza et al. 1998). They deposit a single egg in each hole. Each egg batch can be identified by the presence of several holes in close proximity, placed in a spiral formation. The egg batch results in a multichambered gall (Bondar 1923, Lima 1956). Souza et al. (1998) sampled 175 egg batches from 48 individual plants in January 1992 at the Serra do Cipó, Minas Gerais State, Brazil. They found a very large variation in the number of eggs per egg batch (range 3 – 181) (Table 1). Nevertheless, not all eggs develop to hatch. The hatching success and gall development depend on plant responses (see below).

Table 1   Variable traits of *Collabismus clitellae*, a shoot galler weevil on *Solanum lycocarpum* in SE Brazil.

Trait	n	mean ± SD	range
Number of eggs per egg batch [a]	175	29.3 ± 21.5	3 – 181
Gall volume (cm³) [a]	385	21.7 ± 24.6	0.3 – 140
Pupal weight (mg) [b]	274	108.0 ± 37.1	16 – 176
Adult weight (mg) [b]	124	87.1 ± 21.6	13 – 127

[a] data from Souza et al. (1998)
[b] data from Souza et al. (2001)

Larvae feed on the galled tissues in the chambers from January to July. Gall growth ends in April, but larvae continue to grow and feed until pupation (May to July). In July, adults can still be found in the chambers, but by September and October most of them emerge, matching the end of the drought season and the beginning of the rainy season. Adults are most abundant from October to December, when they can be found feeding on new stems, flower buds and flowers of the host plant. There is a large variation in the time spent by each stage of the weevil to complete life cycle development. This is due to the variation in the number of eggs laid as a function of individual variation in female egg-laying behavior. Pupal and adult mass varies with developmental time, which in turn depends on insect density within the galls (Souza et al. 2001).

Availability of Oviposition Sites and its Influence on Gall Development

We marked 55 egg batches in 1992 at the same site studied by Souza et al. (2001). We evaluated the effects of plant height, stem diameter, and number of eggs laid on the success of gall formation. We used stepwise multiple regression to evaluate linear effects of the number of eggs, and linear and quadratic effects of plant height and stem diameter on the number of chambers formed. We expected a quadratic trend because very young (short) plants would not be able to support large galls, and tall and old plants would have high abortion rates due to the plant's physical and chemical defenses (Souza et al. 2001). The same was expected in relation to stem diameter: the maximum success of gall development was expected on shoots of intermediate sizes. The only significant effects were the linear trends for stem diameter and plant height, and the quadratic effect of plant height (Table 2). Thus, the number of eggs laid had no relationship with gall developmental success but, as expected, galls developed better on plants of intermediate height. The linear effect of stem diameter was negative: there was a poor gall development in thicker shoots. This pattern is possibly due to the fact that the observed egg batches were located only on thicker shoots if one compares the mean diameter of shoots used by the galler. If we had observed development of egg batches on thinner shoots, the quadratic effect of stem diameter could be detected instead of the descendent part of the quadratic curve only. Thus, samples of host plant populations presenting more thinner shoots would be necessary to test if there is a quadratic relationship between stem diameter and gall developmental success.

Table 2 Stepwise multiple regression analysis on factors influencing the developmental success of *C. clitellae* galls, measured by the number of chambers formed inside each gall. $R^2 = 0.99$

Effect	Coefficient	SE	t	P
Constant	−3.334	0.121	−27.5	< 0.001
Stem diameter (linear)	−2.257	0.010	−223.7	< 0.001
Stem diameter (quadratic)			0.475	0.411
Plant height (linear)	1.516	0.125	12.1	< 0.001
Plant height (quadratic)	−0.233	0.032	−7.21	< 0.001
Number of eggs			0.890	0.378

As a result of variation in individual female egg-laying behaviour, galls are not randomly distributed among and within their host plants. Souza et al. (1998) sampled 385 galls from 97 individual (age = 16-18 months) *S. lycocarpum* from single small shoots (<10 cm to 1.6 m from the ground). Individuals up to 3 m in height were also found and sampled in the study area. Galls were distributed on plants ranging from 40 to 200 cm high, with more galls (42%) on plants 80-120 cm tall than expected by chance and only 3% being recorded only on higher than 160 cm plants. Within host plants, 56% were found in basal and 42% in the median parts of the plants.

Nevertheless, 78% of the larger galls ($> 40\ \mathrm{cm}^3$) occurred on the basal stem of the plants, indicating that gall volume is influenced by an interaction between gall relative height and plant height. The diameter of the stems on which galls developed also varied (CV = 0.33), even when stems at similar heights were compared in the same plant. For example, the stem diameters used in the basal part of the host plants varied from 0.47 to 2.83 cm.

Plasticity of Gall Size in Relation to Environmental Factors

Gall volume was highly variable, ranging from 0.34 to 139.8 cm³ (Table 1). This variation represents both the variation in stem diameter and the variation in the number of chambers; the latter ranged from 1 to 70 chambers per gall. Sixty five percent of 385 galls were smaller than 20 cm³ and had 1 to 8 chambers. The large galls ($> 40\ \mathrm{cm}^3$) had up to 68 chambers. Variation in gall size could result from the control of gall size or abortion rate by differential responses of the host plant or by the number of eggs laid by the female. The effectiveness of gall induction on *S. lycocarpum* declines with plant age, constraining gall distribution to smaller plants (from 0.40 to 1.60 m in height). In these smaller plants, gall volume was significantly influenced only by shoot diameter and by the number of chambers (Souza et al. 2001). Thus, larger galls resulted from a combination of more chambers and thicker shoots, which were found mainly in the basal part of the plants (Souza et al. 1998, 2001).

The influence of more chambers and thicker shoots on gall size indicates that the availability of resources per larva may also vary with stem diameter and number of chambers. Within the range of egg-laying sites within and among plants involved in gall development, galls on thinner shoots become smaller than those located on thicker shoots, even though they have similar number of chambers. These smaller galls produce smaller chambers, resulting in lower availability of resources per larva. In the same way, galls of similar volume may have variable number of chambers (Souza et al. 2001), and galls with higher number of chambers should present smaller chamber mean sizes.

Reaction Norms: Developmental Times and Body Sizes

The reduction in chamber sizes implies that the amount of food resources per larva should be lower than for developing larvae in larger chambers. In 1992, we sampled 52 galls in the same population described above. The length and maximum diameter of each chamber was measured and its volume estimated assuming that chamber form was similar to an ellipsoid. The individual from each chamber was weighed and categorized as larva, pupa or adult. Individual weight increased with chamber volume for larvae (y = 19.9 + 411.1 lnx, n = 100, r² = 0.63, p < 0.001), pupae (y = 36.7 + 372.1 lnx, n = 118, r² = 0.79, p < 0.001), and adults (y = 48.9 + 186.7 lnx, n = 98, r² = 0.32, p < 0.001). The rates of increment in insect weight with chamber sizes were smaller at larger chamber sizes. Thus, the size of individuals tends to stabilize at a maximum value (Fig. 1). The relationship between chamber volume and final adult sizes was weaker (only 32 percent of the variation was explained by chamber size), suggesting that other factors involved in metamorphosis could also influence adult final sizes. Nevertheless, the strong relationships found for all developmental stages indicate that chamber size probably reflects the availability of resources for these insects.

The increase in the number of chambers per gall volume leads to a reduction in mean size of chambers per gall, resulting in a limitation of food resources for weevil development. In this study, galls represented a range of different densities that influences individuals' development resulting in a huge variation in *C. clitellae* individual sizes. Indeed, Souza et al. (2001)

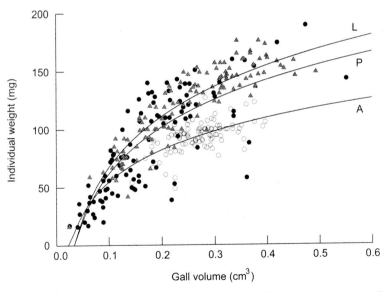

Fig. 1 Relationship between individual weights of larvae (L, solid circles), pupae (P, gray triangles), and adults (A, open circles) of *C. clitellae*.

found a huge variation in the sizes of larvae, pupae, and adults from a sample of 385 galls (Table 1). This variation fitted well to a density-dependence model, as a result of intraspecific competition for host plant resources. Thus, variation in body sizes follows a reaction norm, the final mean sizes of adults decreasing with resource limitation, and smaller adults developing in smaller chamber galls (Fig. 2). The regression lines for larvae, pupae and adults were similar to each other (test for parallelism: $F_{2,308}$ = 1.09, p > 0.3; test for heights: $F_{2,308}$ = 1.51, p > 0.2), indicating that the reduction in the amount of resources *per capita* with the increase in densities led to a similar weight loss to all stages in the weevil life cycle.

On the other hand, the percent of variation in individual weights explained by gall densities were 26 percent for larvae, 51 percent for pupae, and 70 percent for adults (Fig. 2). The scatter of the points is much greater for larvae than for pupae, and in the latter relative to adults. This means that resource limitation may have greater effects as individuals approach the final developmental stages. Thus, it is possible that individuals with weights much below the average predicted by the model die. On the other hand, individuals much above the average predicted by the model could lose weight and adjust its size to the actual density conditions and, thus, to the availability of resources imposed by plant responses.

Final adult size is related to development time, since larvae that spend more time feeding can become larger. Souza et al. (2001) noted that galls collected at the same time contained individuals at distinct stages. For example, galls collected in May contained both larvae and pupae inside the same gall, although they originated from the same egg batch. In a similar way, galls sampled in June and July sometimes had larvae, pupae and adults inside the same gall. The timing of metamorphosis to a different stage influenced strongly the size of individuals, resulting in significant differences in body weights from May to July. Earlier metamorphosed pupae and adults were smaller than those which metamorphosed later. *Collabismus clitellae* seems to respond to variations in environmental conditions by metamorphosing earlier when resource are limited. Indeed a significant relationship between gall density and likelihood of metamorphosis was found by Souza et al. (2001). Larvae in higher density galls had a greater probability of turning into pupae earlier than larvae in less dense galls, which had larger mean chamber size. Similarly, developing pupae in higher density galls had a greater probability of becoming adults earlier. Thus, the increase in density inside galls resulted in a reduction of the time of development due to resource limitation, the largest individuals being those which had more time to complete their development.

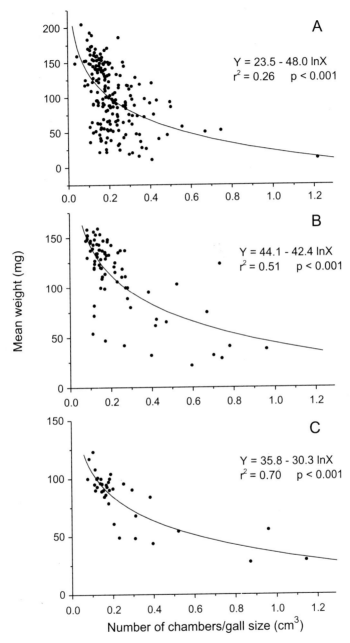

Fig. 2   Relationship between mean weight of larvae (A), pupae (B), and adults (C) of *C. clitellae* and gall larval density (number of individuals per cm³ of gall tissue). (C) was modified from Souza et al. (2001). Results of least-squares regressions are also shown.

Direct and Indirect Effects on C. clitellae Body Size

The mechanism of adjustment of body sizes to different densities described above depends on characteristics of the populations of the host plants and on variation among and within plants to gall induction by the weevil. To evaluate the causal effects of these characteristics on adult final weights, we made a path analysis (Hair et al. 1998). We expected that the environmental conditions determined by characteristics that varied among and within plants influenced variation among galls, which in turn would determine the size of adult weevils. Variation among individual plants such as age, estimated through plant height, and among sites within plants such as gall relative height and shoot diameter used by the galler, are the possible sources of variation in the response of plants to gall formation, determining gall characteristics such as gall size (volume) and number of chambers. Both gall volume and number of chambers would influence the final sizes (weights) of adult insects, by determining the amount of resources available to the larvae.

Our model had a good fit with the observed covariance matrix using maximum likelihood estimation (p > 0.45; see Fig. 3). There were significant influences of plant height and gall relative height on the number of chambers formed, and significant effects of stem diameter and number of chambers on gall volume. The effect of plant height on the number of chambers was similar to that found for another dataset, in which an indirect

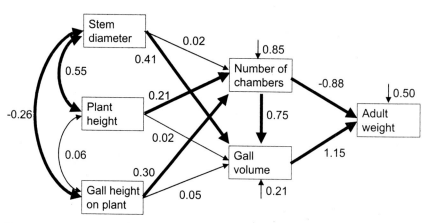

Fig. 3 Path diagram showing causal influences of plant and gall characteristics on adult body weight of *Collabismus clitellae*. Numbers are path coefficients, thick arrows denote significant paths (p < 0.05), thin arrows denote non-significant paths (p > 0.05). The model fit was significant ($\chi^2 = 2.42$, df = 3, p = 0.490).

effect through the number of eggs laid was also detected (Souza et al. 2001). Of all plant characteristics evaluated, stem diameter was positively correlated with plant height and negatively with gall relative height. Galls found in more apical positions were located on thinner shoots, and thus had smaller volumes. Gall size seems to depend also on the number of chambers formed inside it. The path analysis showed that both the number of chambers and the diameter of the shoot, where the eggs were deposited, explained 79% of the variation in gall volume. On the other hand, the number of chambers formed was influenced by other characteristics, such as plant height and gall relative height. However, both variables together explained only 15 percent of the variation in the number of chambers.

Final adult sizes were strongly influenced by a combination of gall volume and number of chambers; larger adults found on larger, few-chambered galls. The direct effects of both variables explained 50% of the variation in the final weights of C. clitellae adults, a value smaller than that recorded by Souza et al. (2001) for the direct relationship between adult weights and density inside galls (70%; see Fig. 3). Density is a better estimator of the availability of resources for larval growth, although other factors could indirectly influence adult sizes. For example, differences in nutrient input to galls situated on different plant positions could be found due to competition between or sinks within the plant (e.g., Larson and Whitham 1997), even if the galls had similar densities.

The large variation in final adult sizes of C. clitellae may thus depend on the number of eggs laid, and to site choice by laying females, because characteristics among and within plants have influence on different parts of the weevil's life-cycle that may influence the availability of resources to the offspring. The main factors that influence the fecundity of C. clitellae are shown in Fig. 4. The "internal" factors are related to the development of the life history of this insect galler, which is directly and indirectly influenced by both plant characteristics and mortality factors such as wind and natural enemies (Souza et al. 1998). Availability of oviposition sites is restricted by the abundance of adequate plant parts that respond to gall formation, whereas mortality factors may restrict even more the sites favorable to the success in adult formation. For example, larger galls that tend to produce larger individuals are attacked more often by fungi, possibly transported to the chamber by a species of Cerambycidae (Coleoptera) (Souza et al. 1998). These authors showed also that large-sized galls in the more apical parts of the host plant have more probability of being attacked by woodpeckers. Wind is also a source of mortality, since larger egg batches laid on thinner shoots increase the probability of stem breakage at the site of oviposition,

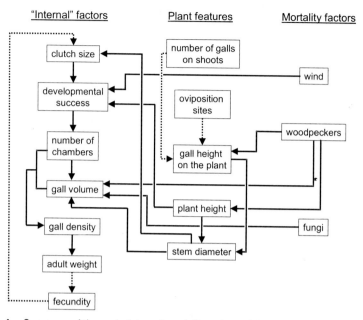

Fig. 4   Summary of the main interactions influencing adult weights and fecundities of *C. clitellae*. Dotted lines are proposed relationships; all other lines are relationships determined in this work and in Souza et al. (1998, 2001).

resulting in a failure of gall formation and offspring death. Parasitoids could also be an important mortality factor to galls, and their probability and success of attack are frequently related to some gall characteristics or to characteristics of the environment where the galls are located, such as plant characteristics or abiotic conditions (Craig et al. 1990, Tscharntke 1992, Weis and Kapelinski 1994, Cronin and Abrahamson 2001). We also noted two species of hymenopteran parasitoids, but they occurred in extremely low frequency (Souza et al. 1998), and thus were not included in Fig. 4. However, it is possible that in other populations or years, the intensity of attack by parasitoids could be a limiting factor to both maximum gall size and its distribution among and within host plants.

Phenotypic Plasticity as a Strategy to Increase the Availability of Resources

The availability of favorable oviposition sites can be an important limiting factor to the population growth of *C. clitellae*. Souza et al. (2001) suggested that phenotypic plasticity in this species would be a feasible strategy allowing an increase in the amount of sites that could be used for oviposition, even though adult individuals with smaller body sizes would be produced at lower quality sites. An advantage is that the offspring would have better chances to survive under high-density conditions, in the case of low availability of high quality sites. Further, the increase in the probability of attack by natural enemies on larger galls could exert a selective pressure towards smaller galls. As mentioned above, the formation of smaller galls depends mainly on the plant response to induction by the galler. To test the hypothesis that the ability to form variable body sizes increases the amount of available sites for oviposition and gall development, we combined in a Principal Component Analysis (PCA) three of the main plant characteristics that determine the performance of *C. clitellae*. The variables were plant height, gall relative height on the plant, and diameter of the stem, and the data were derived from the 385 galls sampled from May to July in 1991. The first two PCA axes explained 80% of the variation, whereas 46% of variance was explained by the first axis only. This component was positively correlated with stem diameter (0.839) and plant height (0.476), and negatively correlated with gall relative height (-0.662). The distribution frequency of large individuals, comprising 25% of the largest individuals, differed from the distribution of all galls in the first axis, here considered as sites available for gall formation (G test, $G = 18.6$, df $= 6$, $p < 0.01$) (Fig. 5). This suggests an increase in the availability of sites through a large variation in body sizes of the offspring. Thus, the plasticity in gall sizes, developmental times, and resulting body sizes of adult weevils may overcompensate in fitness a possible lower individual fecundity of the offspring due to their smaller body sizes. It would increase the survival chances of a greater number of descendants in conditions of high individuals density in relation to the availability of egg-laying sites.

Concluding Remarks

Galling insects commonly present variation in traits such as developmental time and body size, which in turn are strongly related to resource availability. The resources available for larval growth are determined by the response of the host plant and the presence of other gallers, either of the same or other species. The resource base for each individual is determined by estimating the amount of plant tissue available for each growing larva and the susceptibility of the host plant to gall formation. In our study, we showed that the norm of reaction for *C. clitellae* body size varies with gall density. When defining density as the number of individuals per unit of gall volume,

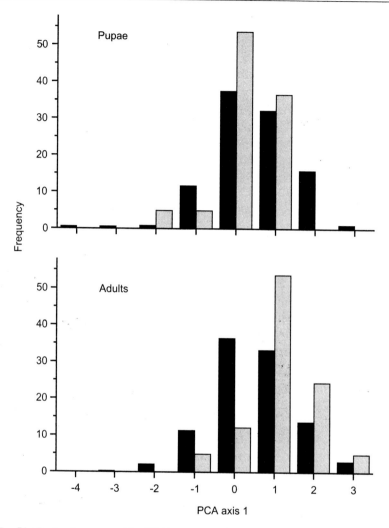

Fig. 5 Distribution frequency of the 25% largest pupae and adults (gray bars) and total galls formed on *S. lycocarpum* in relation to the first PCA axis. Component loadings were: stem diameter = 0.839, plant height = 0.476, gall relative height = -0.662.

we determined how an increase in crowding reduces the amount of resources per individual, resulting in smaller body sizes.

For other systems, it is possible that the use of approaches that explicitly consider the ability of galling insects to vary plastically their sizes and other reproductive traits could allow a better comprehension of the factors that determine the distribution of these insects. The resulting increase in the range of oviposition sites to use suboptimal sites among plant parts or individuals of the same host species can lead to the formation of adults with a similar fitness as those developing in optimal sites, but in low frequency in the population or overexploited by high densities of insects. On the other hand, understanding plastic responses to environmental variation and its influence on insect performance can provide important insights to understand the ecology of gall makers and their shifts to new hosts. The performance of these insects in different plant species, as a result of phenotypic plasticity, could then contribute to the formation of distinct races, and ultimately species.

Acknowledgements

We are grateful for critical discussions of this work with H. R. Pimenta (*in memorian*). We thank the Parque Nacional da Serra do Cipó and Universidade Federal de Minas Gerais for logistical support. A. L. T. Souza and R. P. Martins thank the Conselho Nacional de Pesquisa for scolarships.

References

Abrahamson, W. G. and A. E. Weis. 1987. Nutritional ecology of arthropod gall makers, pp. 235-258. *In* J. R. Slanskey and J. G. Rodriguez [eds] 1987. Nutritional ecology of insects, mites, spiders and related invertebrates. Wiley, New York, USA.

Abrahamson, W. G. and J. F. Sattler, K. D. McCrea, and A. E. Weis. 1989. Variation in selection pressures on the goldenrod gall fly and the competitive interactions of its natural enemies. Oecologia, 79: 15-22.

Abrahamson, W. G. and M. D. Eubanks, C. P. Blair. and A. V. Whipple. 2001. Gall flies, inquilines, and goldenrods: A model for host-race formation and sympatric speciation. American Zoologist, 41: 928-938.

Abrahamson, W. G. and M. D. Hunter, G. Melika, and P. W. Price. 2003. Cynipid gall-wasp communities correlate with oak chemistry. Journal of Chemical Ecology, 29: 208-223.

Aguilar, J. M. and W. J. Boecklen. 1992. Patterns of herbivory in the *Quercus grisea* X *Quercus gambelii* species complex. Oikos, 64: 498-504.

Akimoto, S. 1998. Heterogeneous selective pressures on egg-hatching time and the maintenance of its genetic variance in a *Tetraneura* gall-forming aphid. Ecological Entomology, 23: 229-237.

Akimoto, S. and Y. Yamaguchi. 1994. Phenotypic selection on the process of gall formation of a *Tetraneura* aphid (Pemphigidae). Journal of Animal Ecology, 63: 727-738.

Anderson, S. and K. D. McCrea, W. G. Abrahamson, and L. M. Hartzel. 1989. Host genotypic choice by the gallmaker *Eurosta solidaginis* (Diptera: Tephritidae). Ecology, 70: 1048-1054.

Austin, A. D. and P. C. Dangerfield. 1998. Biology of *Mesostoa kerri* (Insecta : Hymenoptera: Braconidae: Mesostoinae), an endemic Australian wasp that causes stem galls on *Banksia marginata*. Australian Journal of Botany, 46: 559-569.

Björkman, C. 1998. Opposite, linear and non-linear effects of plant stress on a galling aphid. Scandinavian Journal of Forest Research, 13: 177-183.

Björkman, C. 2000. Interactive effects of host resistance and drought stress on the performance of a gall-making aphid living on Norway spruce. Oecologia, 123: 223-231.

Bondar, G. 1923. Biologia do gênero *Collabismus*. Archivos da Escola Superior de Agricultura e Medicina Veterinária, 7: 23-25.

Brown, J. M. and W. G. Abrahamson, R. A. Packer, and P. A. Way. 1995. The role of natural-enemy escape in a gallmaker host-plant shift. Oecologia, 104: 52-60.

Burstein, M. and D. Wool. 1993. Gall aphids do not select optimal galling sites (*Smynthuroides betae*; Pemphigidae). Ecological Entomology, 18: 155-164.

Carvalho, L. A. F. 1985. Flora fanerogâmica da Reserva do Parque Estadual das Fontes do Ipiranga (São Paulo, Brasil). Hoehnea, 12: 67-85.

Clancy, K. M., P. W. Price, and C. F. Sacchi. 1993. Is leaf size important for a leaf-galling sawfly (Hymenoptera, Tenthredinidae). Environmental Entomology, 22: 116-126.

Confer, J. L. and P. Paicos. 1985. Downy woodpecker predation at goldenrod galls. Journal of Field Ornithology, 56: 56-64.

Cornelissen, T. G. and G. W. Fernandes. 2001. Patterns of attack by herbivores on the tropical shrub *Bauhinia brevipes* (Leguminosae): Vigour or chance? European Journal of Entomology, 98: 37-40.

Craig, T. P., P. W. Price, and J. K. Itami. 1986. Resource regulation by a stem-galling sawfly in the arroyo willow. Ecology, 67: 419-425.

Craig, T. P., J. K. Itamy, and P. W. Price. 1989. A strong relationship between oviposition preference and larval performance in a shoot-galling sawfly. Ecology, 70: 1691-1699.

Craig, T. P., P. W. Price, and J. K. Itami. 1990. Intraspecific competition and facilitation by a shoot-galling sawfly. Journal of Animal Ecology, 59: 147-159.

Craig, T. P., J. D. Horner, and J. K. Itami. 2001. Genetics, experience, and host-plant preference in *Eurosta solidaginis*: Implications for host shifts and speciation. Evolution, 55: 773-782.

Craig, T. P. and W. G. Abrahamson, J. K. Itami and J. D. Horner. 1999. Oviposition preference and offspring performance of *Eurosta solidaginis* on genotypes of *Solidago altissima*. Oikos, 86: 119-128.

Craig, T. P. and J. K. Itami, C. Shantz, W. G. Abrahamson, J. D. Horner, and J. V. Craig. 2000. The influence of host plant variation and intraspecific competition on oviposition preference and offspring performance in the host races of *Eurosta solidaginis*. Ecological Entomology, 25: 1-12.

Cronin, J. T. and W. G. Abrahamson. 1999. Host-plant genotype and other herbivores influence goldenrod stem galler preference and performance. Oecologia, 121: 392-404.

Cronin, J. T. and W. G. Abrahamson. 2001. Goldenrod stem galler preference and performance: Effects of multiple herbivores and plant genotypes. Oecologia, 127: 87-96.

De Bruyn, L. 1994. Life history strategies of three gall-forming flies tied to natural variation in growth of *Phragmites australis*, pp 56-72. In P. W. Price, W. J. Mattson, and Y. N. Baranchikov [eds] 1994. The ecology and evolution of gall-forming insects. USDA Forest Service Technical Report NC-174, USDA Forest Service, St Paul, MN, USA.

Desouhant, E., D. Debouzie, and F. Menu. 1998. Oviposition pattern of phytophagous insects: on the importance of host population heterogeneity. Occologia, 114: 382-388.

Eliason, E. A. and D. A. Potter. 2000. Budburst phenology, plant vigor, and host genotype effects on the leaf-galling generation of *Callirhytis cornigera* (Hymenoptera: Cynipidae) on pin oak. Environmental Entomology, 29: 1199-1207.

Fay, P. A., R. W. Preszler and T. G. Whitham. 1996. The functional resource of a gall-forming adelgid. Oecologia, 105: 199-204.

Fernandes,G. W. 1990. Hypersensitivity: a neglected plant resistance mechanism against insect herbivores. Environmental Entomology, 19: 1173-1182.

Fernandes, G. W. 1998. Hypersensitivity as a phenotypic basis of plant induced resistance against a galling insect (Diptera: Cecidomyiidae). Environmental Entomology, 27: 260-267.

Fernandes, G. W. and D. Negreiros. 2001. The occurrence and effectiveness of hypersensitive reaction against galling herbivores across host taxa. Ecological Entomology, 26: 46-55.

Floate, K.D., G. W. Fernandes, and J. A. Nilsson. 1996. Distinguishing intrapopulational categories of plants by their insect faunas: galls on rabbitbush. Oecologia, 105: 221–229.

Freese, G. and H. Zwölfer. 1996. The problem of optimal clutch size in a tritrophic system: The oviposition strategy of the thistle gallfly *Urophora cardui* (Diptera, Tephritidae). Oecologia, 108: 293-302.

Fritz, R. S., C. M. Nichols-Orians, and S. J. Brunsfeld. 1994. Interspecific hybridization of plants and resistance to herbivores: hypotheses, genetics, and variable responses in a diverse herbivore community. Oecologia, 97: 106–117.

Fritz, R. S., B. A. Crabb, and C. G. Hochwender. 2000. Preference and performance of a gall-inducing sawfly: a test of the plant vigor hypothesis. Oikos, 89: 555-563.

Glynn, C. and S. Larsson. 1994. Gall initiation success and fecundity of *Dasineura marginemtorquens* on variable *Salix viminalis* host plants. Entomologia Experimentalis et Applicata, 73:11-17.

Graham, J. H., E. D. McArthur, and D. C. Freeman. 2001. Narrow hybrid zone between two subspecies of big sagebrush (*Artemisia tridentata* : Asteraceae) - XII. Galls on sagebrush in a reciprocal transplant garden. Oecologia, 126: 239-246.

Hair, J. F., Jr. and R. E. Anderson, R. L. Tatham, and W. C. Black. 1998. Multivariate data analysis, 5th edn. Prentice-Hall, New Jersey, USA.

Hartley, S. E. and J. H. Lawton. 1992. Host-plant manipulation by gall-insects - a test of the nutrition hypothesis. Journal of Animal Ecology, 61: 113-119.

Hawkins, B. A., H. V. Cornell and M. E. Hochberg. 1997. Predators, parasitoids, and pathogens as mortality agents in phytophagous insect populations. Ecology, 78: 2145-2152.

Hellqvist, S. 2001. Biotypes of *Dasineura tetensi*, differing in ability to gall and develop on black currant genotypes. Entomologia Experimentalis et Applicata, 98: 85-94.

Hellqvist, S. and S. Larsson. 1998. Host acceptance and larval development of the gall midge *Dasineura tetensi* (Diptera, Cecidomyiidae) on resistant and susceptible black currant. Entomologica Fennica, 9: 95-102.

Honek, A. 1993. Intraspecific variation in body size and fecundity in insects – a general relationship. Oikos, 66: 483-492.

Horner, J. D. and W. G. Abrahamson. 1992. Influence of plant genotype and environment on oviposition preference and offspring survival in a gallmaking herbivore. Oecologia, 90: 323-332.

Horner, J. D. and W. G. Abrahamson. 1999. Influence of plant genotype and early-season water deficits on oviposition preference and offspring performance in *Eurosta solidaginis* (Diptera: Tephritidae). American Midland Naturalist, 142: 162-172.

Horner J. D., T. P. Craig, and J. K. Itami. 1999. The influence of oviposition phenology on survival in host races of *Eurosta solidaginis*. Entomologia Experimentalis et Applicata, 93: 121-129.

Inbar, M., A. Eshel, and D. Wool. 1995. Interspecific competition among phloem-feeding insects mediated by induced host-plant sinks. Ecology, 76: 1506-1515.

Junichi, Y. 2000. Synchronization of gallers with host plant phenology. Population Ecology, 42: 105-113.

Kearsley M. J. C. & T. G. Whitham. 1989. Developmental changes in resistance to herbivory: implications for individuals and populations. Ecology, 70: 422-434.

Keep, E. 1985. The black currant leaf curling midge, *Dasyneura tetensi* Rübs., its host range, and the inheritance of host resistance. Euphytica, 34: 801-809.

Kimberling, D. N., E. R. Scott, and P. W. Price. 1990. Testing a new hypothesis: plant vigour and phylloxera distribution on wild grape in Arizona. Oecologia, 84: 1-8.

Klingenberg, C. P. and J. R. Spence. 1997. On the role of body size for life-history evolution. Ecological Entomology, 22: 55-68.

Kokkonen, K. 2000. Mixed significance of plant vigor: two species of galling *Pontania* in a hybridizing willow complex. Oikos, 90: 97-106.

Larson, K. C. and T. G. Whitham. 1991. Manipulation of food resources by a gall-forming aphid: the physiology of sink-source interactions. Oecologia, 88: 15-21.

Larson, K. C. and T. G. Whitham. 1997. Competition between gall aphids and natural plant sinks: plant architecture affects resistance to galling. Oecologia, 109: 575-582.

Larsson, S. and D. R. Strong. 1992. Oviposition choice and larval survival of *Dasyneura marginemtorquens* (Diptera: Cecidomyiidae) on resistant and susceptible *Salix viminalis*. Ecological Entomology, 17: 227-232.

Larsson, S. and B. Ekbom. 1995. Oviposition mistakes in herbivorous insects - confusion or a step towards a new host giant. Oikos, 72: 155-160.

Larsson, S., C. Glynn, and S. Höglund. 1995. High oviposition rate of *Dasyneura marginemtorquens* on *Salix viminalis* genotypes unsuitable for offspring survival. Entomologia Experimentalis et Applicata, 77: 263-270.

Lichter, J. P., A. E. Weis, and C. R. Dimick. 1990. Growth and survivorship differences in *Eurosta* (Diptera; Tephritidae) galling sympatric host plants. Environmental Entomology, 19: 972-977.

Lima, A. M. M. C. 1956. Insetos do Brasil 10º Tomo: Coleópteros. Escola Nacional de Agronomia, Série didática nº 12, Rio de Janeiro, RJ, Brazil.

McKinnon, M. L., D. T. Quiring, and E. Bauce. 1999. Influence of tree growth rate, shoot size and foliar chemistry on the abundance and performance of a galling adelgid. Functional Ecology, 13: 859-867.

Nylin, S. and K. Gotthard. 1998. Plasticity in life-history traits. Annual Review of Entomology, 43: 63-83.

Ollerstam, O. and O. Rohfritsch, S. Hoglund, and S. Larsson. 2002. A rapid hypersensitive response associated with resistance in the willow *Salix viminalis* against the gall midge *Dasineura marginemtorquens*. Entomologia Experimentalis et Applicata, 102: 153-162.

Ozaki, K. 1993. Effects of gall volume on survival and fecundity of gall-making aphids *Adelges japonicus* (Homoptera, Adelgidae). Researches on Population Ecology, 35: 273-284.

Ozaki, K. 2000. Insect-plant interations among gall size determinants of adelgids. Ecological Entomology, 25:452-459.

Pires, C. S. S. and P. W. Price. 2000. Patterns of host plant growth and attack and establishment of gall-inducing wasp (Hymenoptera: Cynipidae). Environmental Entomology, 29: 49-54.

Prado, P. I. K. L. and E. M. Vieira. 1999. The interplay between plant traits and herbivore attack: a study of a stem galling midge in the neotropics. Ecological Entomology, 24: 80-88.

Preszler, R. W. and P. W. Price. 1988. Host quality and sawfly populations: a new approach to life table analysis. Ecology, 69: 2012-2020.

Price, P. W. 1991. The plant vigor hypothesis and herbivore attack. Oikos, 62: 244-251.

Price, P. W. and K. M. Clancy. 1986. Interactions among three trophic levels: Gall size and parasitoid attack. Ecology, 67: 1593-1600.

Price, P. W., H. Roininen, and J. Tahvanainen. 1987a. Why does the bud-galling sawfly, *Euura mucronata* attack long shoots? Oecologia, 74:1-6.

Price, P. W., H. Roininen, and J. Tahvanainen. 1987b. Plant age and attack by the bud galler, *Euura mucronata*. Oecologia, 73: 334-337.

Price, P. W., H. Roininen, and J. Tahvanainen. 1997. Willow tree shoot module length and the attack and survival pattern of a shoot-galling sawfly, *Euura atra* (Hymenoptera: Tenthredinidae). Entomological Fennica, 8: 113-119.

Price, P. W., H. Roininen, and T. Ohgushi. 1999. Comparative plant-herbivore interactions involving willows and three gall-inducing sawfly species in the genus *Pontania* (Hymenoptera: Tenthredinidae). Ecoscience, 6: 41-50.

Raman, A. and W. G. Abrahamson. 1995. Morphometric relationships and energy allocation in the apical rosette galls of *Solidago altissima* (Asteraceae) induced by *Rhopalomyia solidaginis* (Diptera, Cecidomyiidae). Environmental Entomology, 24: 635-639.

Rehill, B. J. and J. C. Schultz. 2001. *Hormaphis hamamelidis* and gall size: a test of the plant vigor hypothesis. Oikos, 95: 94-104.

Rehill, B. J. and J. C. Schultz. 2002. Opposing survivorship and fecundity effects of host phenology on the gall-forming aphid *Hormaphis hamamelidis*. Ecological Entomology, 27: 475-483.

Roff, D. A. 1992. The evolution of life histories: theory and analysis. Chapman & Hall, New York, USA.

Rossi, A. M. and P. Stiling, M. V. Cattell, and T. I. Bowdish. 1999. Evidence for host-associated races in a gall-forming midge: trade-offs in potential fecundity. Ecological Entomology, 24: 95-102.

Shibata, E. 2001. Synchronization of shoot elongation in the bamboo *Phyllostachys heterocycla* (Monocotyledoneae: Gramineae) and emergence of the gall maker *Aiolomorphus rhopaloides* (Hymenoptera: Eurytomidae) and its inquiline *Diomorus aiolomorphi* (Hymenoptera: Torymidae). Environmental Entomology, 30: 1098-1102.

Sopow, S. L. and D. T. Quiring. 2001. Is gall size a good indicator of adelgid fitness? Entomologia Experimentalis et Applicata, 99:267-271.

Souza, A. L. T. and G. W. Fernandes, J. E. C. Figueira, and M. O. Tanaka. 1998. Natural history of a gall-inducing weevil *Collabismus clitellae* (Coleoptera: Curculionidae) and some effects on its host plant *Solanum lycocarpum* (Solanaceae) in southeastern Brazil. Annals of the Entomological Society of America, 91: 404-409.

Souza, A. L. T. and M. O. Tanaka, G. W. Fernandes, and J. E. C. Figueira. 2001. Host plant response and phenotypic plasticity of a galling weevil (*Collabismus clitellae*: Curculionidae). Austral Ecology, 26: 173-178.

Stearns, S. C. 1992. The evolution of life histories. Oxford University Press, Oxford, UK.

Stiling, P. and A. M. Rossi. 1996. Complex effects of genotype and environment on insect herbivores and their enemies. Ecology, 77: 2212-2218.

Thompson, J. N. 1988. Evolutionary ecology of the relationship between oviposition, preference and performance of offspring in phytophagous insects. Entomologia Experimentalis et Applicata, 47: 3-14.

Tscharntke, T. 1992. Cascade effects among four trophic levels: bird predation on galls affects density-dependent parasitism. Ecology, 73: 1689-1698.

Walton, R., A. E. Weis, and J. P. Lichter. 1990. Oviposition behavior and response to plant height by *Eurosta solidaginis* Fitch (Diptera: Tephritidae). Annals of the Entomological Society of America, 83: 509-514.

Weis, A. E. 1992. Plant variation and the evolution of phenotypic plasticity in herbivore preference, pp 140-171. *In* R. S. Fritz and E. L. Simms [eds] 1992. Plant resistance to herbivores and pathogens. University of Chicago Press, Chicago, USA.

Weis, A. E. 1996. Variable selection of *Eurosta's* gall size, III: Can an evolutionary response to selection be detected? Journal of Evolutionary Biology, 9: 623-640.

Weis, A. E. and W. G. Abrahamson. 1986. Evolution of host-plant manipulation by gall makers: Ecological and genetic factors in the *Solidago-Eurosta* system. American Naturalist, 127: 681-695.

Weis, A. E. and W. L. Gorman. 1990. Measuring selection on reaction norms: an exploration of the *Eurosta-Solidago* system. Evolution, 44: 820-831.

Weis, A. E. and A. Kapelinski. 1994. Variable selection of *Eurosta's* gall size. II. A path analysis of ecological factors behind selection. Evolution, 48: 734-745.

Weis A. W., R. Walton, and C. L. Crego. 1988. Reactive plant tissue sites and the population biology of gall makers. Annual Review of Entomology, 33: 467-486.

Weis, A. E., W. G. Abrahamson, and M. C. Andersen. 1992. Variable selection of *Eurosta's* gall size, I: The extent and nature of variation in phenotypic selection. Evolution, 46: 1674-1697.

Whitham, T. G. 1978. Habitat selection by *Pemphigus* aphids in response to resource limitation and competition. Ecology 59: 1164-1176

Whitham, T. G. 1980. The theory of habitat selection: examined and extended using *Pemphigus* aphids. American Naturalist, 115: 449-465.

Whitham, T. G., A. G. Williams, and A. M. Robinson. 1984. The variation principle: individual plants as temporal and spatial mosaics of resistance to rapidly evolving pests, pp 15-51. *In* P. W. Price, C. N. Slobodchikoff, and W. S. Gand [eds] 1984. A new ecology: novel approaches to interative systems. Wiley, New York, USA.

Williams, A. G. and T. G. Whitham. 1986. Premature leaf abscission - an induced plant defense against gall aphids. Ecology, 67: 1619-1627.

Woods, J. O. and T. G. Carr, P. W. Price, L. E. Stevens, and N. S. Cobb. 1996. Growth of coyote willow and the attack and survival of a mid-rib galling sawfly, *Euura* sp. Oecologia, 108: 714-722.

Wool, D. 2004. Galling aphids: specialization, biological complexity, and variation. Annual Review of Entomology, 49: 175-192.

Wool, D. and N. Bar-El. 1995. Population ecology of the galling aphid *Forda formicaria* von Heyden in Israel: abundance, demography, and gall structure. Israel Journal of Zoology, 41: 175-192.

Wool, D. and O. Manheim. 1988. The effects of host-plant properties on gall density, gall weight and clone size in the aphid *Aploneura lentisci* (Pass.) (Aphididae: Fordinae) in Israel. Researches on Population Ecology, 30: 227-234.

Species Index

Subject Index